CAMBRIDGE ENGLISH PROSE TEXTS

English Science, Bacon to Newton

CAMBRIDGE ENGLISH PROSE TEXTS

General editor: GRAHAM STOREY

OTHER BOOKS IN THE SERIES
Revolutionary Prose of the English Civil War,
edited by Howard Erskine-Hill & Graham Storey
The Evangelical and Oxford Movements,
edited by Elisabeth Jay
Science and Religion in the Nineteenth Century,
edited by Tess Cosslett
American Colonial Prose: John Smith to Thomas Jefferson,
edited by Mary Ann Radzinowicz
Burke, Paine, Godwin and the Revolution Controversy,
edited by Marilyn Butler
Victorian Criticism of the Novel,
edited by Edwin M. Eigner and George J. Worth
Critics of Capitalism: Victorian Reactions to 'Political Economy',
edited by Elizabeth and Richard Jay

FORTHCOMING
Romantic Critical Essays, edited by David Bromwich

English Science, Bacon to Newton

edited by

BRIAN VICKERS

*Professor of English and
Renaissance Literature
ETH Zürich*

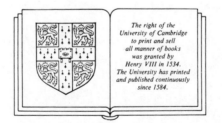

The right of the
University of Cambridge
to print and sell
all manner of books
was granted by
Henry VIII in 1534.
The University has printed
and published continuously
since 1584.

CAMBRIDGE UNIVERSITY PRESS

Cambridge
London   New York   New Rochelle
Melbourne   Sydney

Published by the Press Syndicate of the University of Cambridge
The Pitt Building, Trumpington Street, Cambridge CB2 1RP
32 East 57th Street, New York, NY 10022, USA
10 Stamford Road, Oakleigh, Melbourne 3166, Australia

First published 1987

Printed in Great Britain at
the University Press, Cambridge

British Library cataloguing in publication data

English Science, Bacon to Newton. –
(Cambridge English prose texts)
1. Science – England – History – 17th century
I. Vickers, Brian, 1937–
509'.42   Q127.G4

Library of Congress cataloguing in publication data

English Science, Bacon to Newton.
(Cambridge English prose texts)
Bibliography.
1. Science – Great Britain – History.
I. Vickers, Brian. II. Series.
Q127.G4E54   1987   509.41   86–19265

ISBN 0 521 30408 3 hard covers
ISBN 0 521 31683 9 paperback

CE

Contents

Plates

Acknowledgements

For permission to quote from Isaac Newton's *Correspondence*, ed. H. W. Turnbull, I am grateful to the Syndics of Cambridge University Press; for the right to reproduce Newton's manuscript sketch I am grateful to the Syndics of Cambridge University Library.

In Memoriam
Charles B. Schmitt
1933–1986

Preface

When the publisher invited me to prepare a volume on seventeenth-century English science for this series, my first idea was to cover the period from Bacon to Newton, that is, from the 1620s to the 1670s. After working on this idea for some time I came to realize that it would involve reprinting a number of texts in which Bacon's ideas were repeated and re-formulated but without being taken further, either through the classic Baconian recipe for science, a combination of observations and experiment resulting in the formulation of new scientific laws, or by a critique of Bacon's work from a fresh point of view. There was much discussion of the Baconian plan for science in the 1640s and 50s, in the circle around the Puritan educational reformer Samuel Hartlib; in a group connected with Gresham College, London; and in another group associated with John Wilkins at Wadham College, Oxford (groups whose membership overlapped), discussions and movements admirably documented by Charles Webster.[1] But they yielded very little of what Bacon called light and fruit, new experiments and new discoveries. Nor did they contribute very much to the topic that this series aims to illuminate, the nature of English prose.

On further thought, then, it seemed better to shift the centre of interest to the 1660s, one of the most fruitful decades in the history of English science, when the social and political stability created by the Restoration (see Sprat p. 173 below) encouraged the institutionalizing of science – although, of course, there were still scholars who had a large enough private income to work outside institutions. Starting from the 1660s would enable me to cover the foundation of the Royal Society, and to include the work of its three most prolific and gifted members, Hooke, Boyle, and Newton. (Of the other leading lights, Wren published very little, and John Ray issued his major works in Latin.) Since virtually all these natural philosophers ("scientists" is an anachronism, strictly speaking, but almost unavoidable) expressed their intellectual debts to Bacon, and since Sprat made him the Society's inspiration, the frontispiece to his *History* giving his bust the place of honour beside its royal patron, Charles II, it seemed legitimate to begin my anthology with two excerpts from Bacon himself, one a seminal account of the organization of a scientific research institute, the other a no less important statement of the way in which experiments should be written up. The introduction, accordingly, begins with an account of Bacon's influence on seventeenth-century science. This is history written, as it were, from the Royal Society's own perspective, for Sprat declared that there would have been no better preface to his *History* than "some of [Bacon's] writings".

Focusing on the 1660s had the other advantage of allowing me to include two documents frequently invoked in discussions of the relations between science and language, Sprat's account of the new prose style that the Royal Society had supposedly instituted, and Wilkins's project for abolishing language in favour of an "unambiguous" sign-system. I have not just endorsed received opinion on this issue, for my own analysis of the evidence leads me to question whether Sprat's reform ever took place, and to suggest that his remarks should be understood in more specific contexts involving the incorporation of the Royal Society into the ruling establishment of church and state, and the role of English as an adequate language for science.[2] The prose style of Joseph Glanvill, often described as an instance of these reforms, turns out on closer inspection to be more ambivalent, as the Appendix will show.

For these reasons this anthology begins with Bacon but then moves on to the 1660s and 70s, rather like the Prologue's description of the re-arrangement of history in *Henry V*, "jumping o'er times" in order to "digest" a play. Bacon's influence is seen throughout, not least in the way in which experiments came to be written up. One influential strand in recent history of science has been the recognition that the written-up experiment seldom represents the scientist's experience "raw", with all its false starts and misconceptions, as Kepler's long autobiographical narratives had done. Rather, the archetypal scientific paper arranges its material so as to suggest a frictionless process of formulating a problem, testing it, and producing meaningful results after a reasonable amount of time and effort. Modern scientists have described this as a mythical and misleading process of constructing scientific knowledge by burying failures.[3] Whatever our position on this issue, the work of Bacon, Boyle and Newton was influential in establishing this model for subsequent science. If this anthology had a sub-title it would be "The prose of experiment". That, at any rate, was the tradition I set out to document, fully aware that it is not the whole story of seventeenth-century science. But it was an important part, and more than any other ties up with our wider interest in the course of English prose.

I should like to thank Terry Moore of C.U.P. for his patience and encouragement; Nancy-Jane Thompson for her skilled copy-editing; and my assistant, Dr Margrit Soland, for invaluable help with preparing the manuscript and correcting proofs. For any remaining errors I am alone responsible.

Just as we were going to press news came of the tragically early death of my friend Charles B. Schmitt, a great loss to the History of Science and Renaissance studies. This book is dedicated to his memory.

Note on the texts

Words denoted by an asterisk will be found in the Glossary at the end of the volume.

Introduction

The legacy of Francis Bacon

English science in the seventeenth century acknowledged a great debt to the work of Francis Bacon (1561–1626). Today we value Bacon not so much as a scientist – he made no major discovery, formulated no scientific law, performed few original experiments – but as a propagandist for science who urged that "natural philosophy" be given a new importance in human affairs and be organized on a new plan. From his early masques and "devices" of the 1590s,[1] to the *Two Bookes of Francis Bacon. Of the Proficience and Advancement of Learning, Divine and Humane* (1605), expanded in the Latin translation of 1623 (*De Dignitate et Augmentis Scientiarum, Libros IX*), and the major works of the 1620s, the *Instauratio Magna* (1620), containing his *Novum Organum* or "New Instrument" of scientific method, the *New Atlantis* (1624), and the *Sylva Sylvarum: or a Natural History. In Ten Centuries* (1626), Bacon found time in a busy career in politics and law to formulate a new programme for science.[2] Some of his ideas, such as the attack on medieval scholasticism and its purely philological practice of science, had been expressed by others in the European Renaissance, but he was the first to develop a coherent critique of outmoded science and to suggest practical remedies, while no one else expressed these ideas so forcibly and so eloquently. Yet in the seventeenth century Bacon was actually revered as one of the pioneers in the new philosophy, both as a theorist and an experimenter, his fame spreading throughout Europe.

Bacon's new plan started by freeing the scientist from a reverential attitude towards antiquity. Whereas much medieval science derived from the text of the greatest classical authority, Aristotle, and his many commentators, Bacon exposed the limitations of such reverence in a striking metaphor: "Knowledge is like a water that will never arise again higher than the level from which it fell; and therefore to go beyond Aristotle by the light of Aristotle is to think that a borrowed light can increase the original light from which it is taken" (III, 227, 290; IV, 12). Further, Aristotelian science was a verbal science, passed

I

down in texts and commentaries, relying on a verbal discipline, logic, which can manipulate received ideas but not find out new sciences. "The syllogism consists of propositions, propositions consist of words, words are symbols of notions. Therefore if the notions themselves (which is the root of the matter) are confused and over-hastily abstracted from the facts, there can be no firmness in the superstructure" (IV, 49). The remedy, according to Bacon, was to start from a fresh examination of reality in all its detail, not rejecting, as earlier theorists had done, certain topics as trivial or sordid, and on the basis of observation and experiment gradually build up axioms or scientific laws of increasing generality and universality. From this level one could return to works, or specific man-made trans-formations of nature (IV, 24ff., 92ff.).

"All depends on keeping the eye steadily fixed upon the facts of nature", Bacon wrote, "and so receiving their images simply as they are" (IV, 32). The chief obstacle to this ideal of fidelity to experience is the propensity of the human mind to be deceived by phantoms or fallacies, what he called human "idols". Drawing on the original sense of the Greek word *eidolon*, "an image in the mind, a mental fiction", Bacon distinguished four types of misconception. The "Idols of the Tribe" concern human perception, which far from being absolutely objective and reliable varies according to the measure of the individual. The "Idols of the Cave" refer to each individual's "cave or den", those factors in his personality, psyche, or education that cause him to "refract or discolour the light of nature". Human communication is through language, and the "Idols of the Market-place" are the distortions caused by "the ill and unfit choice of words", which can obstruct and deceive the understanding itself. Lastly the "Idols of the Theatre" are all extant philosophical systems, which are "but so many stage-plays, representing worlds of their own creation" in an unreal and fictitious manner (IV, 54–5). The deficiencies inherent in human perception and communication must be overcome before the mind can enter into a true relation with the universe, from which marriage "let us hope ... there may spring helps to man, and a line and race of inventions that may in some degree subdue and overcome the necessities and miseries of humanity" (IV, 27). Bacon's whole system is directed towards producing inventions and processes which will be useful to man, not in the debased modern sense of "utilitarian-ism" (in which material profit is pursued without concern for ethics), but in the older traditions of the *vita activa* or involvement in society for the benefit of others (an idea that goes back to Plato and Cicero), and Christian charity. In Bacon "usefulness" is equivalent to "philanthropy".[3] The scientist in the *New Atlantis* is described as having "an aspect as if he pitied men" (III, 154).

Bacon's system was designed to unite "the empirical and the rational faculty" (IV, 19), that is, the powers of observation and theory. Once the mind has become a *tabula abrasa* (IV, 27), it could begin to classify the sciences, find where research is needed, and then evolve "Directions concerning the Interpretation of Nature", the new logic worked out in most detail in the *Novum Organum*. Where traditional logic operated deductively, from first principles down to ever more minute distinctions, Bacon would reverse the process by using induction, ascending from a host of specific instances to general propositions. At this stage experiments would be used, but not in a hasty way, nor designed for a swift material application. The scientist must use "experiments of light" to induce new axioms, but he must also proceed deductively to generate "experiments of fruit" that will produce new works (IV, 71, 95). Experimentation must not be random, rather, experiments must be "skilfully and artificially devised for the express purpose of determining the point in question" (IV, 26). In order to prevent an over-development of the theoretical phase recourse must be constantly made to "the Phenomena of the Universe; or a Natural and Experimental History for the foundation of Philosophy" (IV, 22). Bacon proposed the compilation of a "History" – in the sense of a systematic analysis, not necessarily diachronic – of all the phenomena of nature, adding a list of over a hundred suggested titles to his *Preparative* (IV, 265–71). In the frantically busy last years of his life he produced a specimen *Historia Naturalis et Experimentalis*, with histories of the winds (II, 7–78), of life and death (II, 101–226), and of the dense and rare (II, 241–305).

Yet Bacon was not a mere fact-collector. This amassing of concrete observation and experimental results was to form the basis of the last three stages of his *Great Instauration*, his restoration or renewal of science. Next would come the "Ladder of the Intellect", moving from specific phenomena up the ladder of axioms to the "summary law of nature" (IV, 31). This would be followed by "Anticipations of the New Philosophy" (IV, 31–2), tentative generalizations – in effect, hypotheses – to be tested by further observation and analysis. Finally, the Baconian scientist would reach "the New Philosophy or Active Science", the complete system ascending through axioms to natural laws and descending to works (IV, 32–3). Since Bacon only completed fragments of the whole scheme, mostly the earlier parts concerned with the collection of data, he has been wrongly accused of advocating a random amassing of facts: but his plan involved theory and hypothesis as an organic part of the enterprise. The whole system would be a work of many years and of many hands, for Bacon foresaw that large scientific enterprises would only be possible given collaboration and co-ordination. In his Utopian fable, *New Atlantis* (written *c.*1624) he outlined a whole scientific research institute, with a fully worked out division of responsibilities. He also saw that at many levels science would depend not on geniuses but on people of ordinary ability, trained to perform specific but limited tasks. Unlike the great individualists of the scientific revolution – Galileo, Kepler, Descartes, Newton – Bacon was thinking in terms of institutions that would channel scientific research and then apply it to the benefit of society at large. His achievement was lesser than theirs, of course, but he did redirect attention to several important aspects of science.

The next generations certainly gave him all credit for his programme and for the vision of ultimate success that he so memorably formulated. As a master of rhetoric Bacon used his skills of persuasion to stimulate his readers to the practice of science in the conviction of its value to humanity.[4] Although it has been claimed that Bacon only influenced minor and eccentric figures,[5] in fact many of the leading English scientists between 1630 and 1680 acknowledged the inspiration they had drawn from him.[6] William Petty described Bacon as "the Master-builder" of the revival of science, praising his "most excellent specimen" of a history of nature, and his "exact and judicious Catalogue" of particulars.[7] Petty's plan to compile a History of Trades (that is, Crafts) – a project that occupied many writers, from Boyle to Defoe – owed its inspiration to Bacon, as did his project to found a "mechanical and medical college". Sir Thomas Browne's *Pseudodoxia Epidemica* (1646) offered a catalogue of "vulgar errors" as proposed by Bacon (III, 221; IV, 169, 295). Robert Hooke's *General Scheme, or Idea of the Present State of Natural Philosophy, and how its defects may be Remedied by a Methodical Proceeding in the making Experiments and collecting Observations. Whereby to Compile a Natural History as the Solid Basis for the Superstructure of True Philosophy*,[8] is a fantasia on Baconian themes, a digest of the *Novum Organum*. Henry Power, one of the pioneers in microscopy, described Bacon as "that Patriark of Experimental Philosophy".[9] Thomas Sydenham, one of the outstanding doctors of the period, who laid the foundation of the practice of clinical medicine in England, described Bacon as "that great genius of rational nature".[10] Robert Boyle, second only to Newton as a scientist, collected material for a continuation of Bacon's *Sylva Sylvarum*,[11] unstintingly praised "the illustrious Lord *Verulam*, one of the most judicious naturalists

that our age can boast", "that great and solid philosopher", "one of
the first and greatest experimental philosophers of our age",[12] and
produced an influential work in the Baconian tradition of philanthropy, *Some Considerations touching the Usefulness of Experimental
Natural Philosophy* (part I: Oxford, 1663; part II: Oxford, 1671).[13]
Bacon remained a figure of inspiration and emulation.

The Royal Society

When the Royal Society was officially founded in 1662, it united
several strands of scientific activity. One group of scientists had
started meeting in London in 1645, including John Wallis (1616–1703),
the mathematician; John Wilkins (1614–1672), and Jonathan Goddard
(?1617–1675). They sometimes met in Gresham College, a centre for
practical education in the vernacular founded in 1597, which was later
to be the home of the Royal Society for many years. Independent of
this group were the scientists (especially Boyle and Petty) connected
with Samuel Hartlib, the indefatigable Puritan educational reformer,
who poured out an unceasing flow of ideas and projects. Robert
Boyle wrote in 1646–7 of the *"invisible ... philosophical college"*
surrounding Hartlib, its "midwife and nurse", and described his own
studies in "natural philosophy, the mechanics, and husbandry,
according to the principles of our new philosophical college, that
values no knowledge, but as it hath a tendency to use".[14] The
Baconian influence on Hartlib's circle was enormous. His son
Clodius, a chemist, wanted to establish, with Sir Kenelm Digby, "an
universal laboratory ... as may redound, not only to the good of this
island, but also to the health and wealth of all mankind", and in May
1654 Hartlib wrote to Boyle about a visit he had made to Lambeth
Marsh "to see part of that foundation or building, which is designed
for the execution of my lord *Verulam's New Atlantis*".[15] This project,
never realized, was one of a number of contemporary schemes for
setting up scientific research institutes, by Hartlib, Evelyn, Cowley,
and the continuator (probably Robert Hooke) of Bacon's *New
Atlantis* in 1660.[16]

A third group, including some of those who had met in London,
gathered in Oxford following the appointment of Wilkins as Warden
of Wadham College in 1648. They attracted a distinguished group of
younger men: Boyle, Petty, Seth Ward (1617–1689), Thomas Willis
(1621–1675), Richard Lower (1631–1691), and Robert Hooke (1635–
1702). This group joined forces with those scientists who had

continued to meet in London, and on 28 November 1660, in the rooms of Lawrence Rooke, professor of geometry at Gresham College, following a lecture by Christopher Wren, professor of astronomy, they formally constituted themselves into an academy "for the advancement of various parts of learning". Charles II awarded them their charter in 1662, and in the frontispiece to Thomas Sprat's *History of the Royal-Society* (1667) the King shares the honours with the first president, William, Viscount Brouncker, and Francis Bacon, *"Artium Instaurator"*. A key figure in the early years was Henry Oldenburg, Secretary from 1662 until his death in 1677, who carried on Hartlib's tradition of collecting and disseminating information. Bacon, as Paolo Rossi has shown,[17] had made a decisive break with the magical and occult traditions, which kept knowledge secret among the initiated, urging that it was a common good, to be shared by all. (In the *New Atlantis*, however, he recognizes that some scientific discoveries must be restricted, in the national interest: p. 43 below.) The ethos of free circulation of knowledge, so important in the *respublica litterarum* of the seventeenth century, was widely applied to science, as in Sprat's vigorous advocacy of its advantages (pp. 165–6 below). Oldenburg made scientists communicate their discoveries, which he would pass on in correspondence or publish in the Society's own journal, *Philosophical Transactions*, which he edited with enormous industry and enterprise. He instigated the full Baconian programme of natural philosophy, declaring the Society's business to be "in the first place, to scrutinize the whole of Nature and to investigate its activity and powers by means of observations and experiments; and then in course of time to hammer out a more solid philosophy and more ample amenities of civilization". Where Bacon, at the age of 31, had written "I have taken all knowledge to be my province" (VIII, 109), Oldenburg reminded a correspondent "that we have taken to taske the whole Universe, and that we were obliged to doe so by the nature of our dessein".[18]

The Royal Society was not only a clearing-house for information but also a research centre in the sense that its weekly meetings included experiments (for many years performed by that mechanical genius Robert Hooke), which were commented on both orally and in the pages of the *Transactions*. Many scientists associated with the Society continued to work independently, often on private means (unlike present-day institutions it did not receive lavish government support), but still used its meetings to acquire and exchange information. There was no guiding principle for selecting or systematizing experiments, and the accounts of their activities given by Sprat, or recorded in the *Transactions*, reveal a bewildering spread of energy into every possible area, important or trivial. The catholicity of their interest was a weakness in terms of organized, directed research, but a strength in so far as it allowed little of any importance to escape them. Yet their activities were not unanimously approved. Many critics wanted instant proofs of the success of the experimental method, and the question "what have they done?" was one that apologists, such as Sprat, Glanvill, Hooke, and Wallis, attempted to answer.[19] Some of the critics were disgruntled academics from Oxford and Cambridge, hostile to the growth of the new science, or concerned that their own prestige was at stake.[20] Others were conservatives or reactionaries who defended the philosophy of Aristotle as all-sufficient. Although some contemporary historians have championed the opponents, and although the achievements of the Royal Society by 1670 were not as great as might have been hoped, the work of Boyle, Hooke, Ray, Willughby, Wren, and Newton is ample justification for the existence of this institution.

The works represented here have been chosen to illustrate the main course of English science in this period. With the exception of one work by Bacon, originally written in Latin, whose importance was such as to justify the use of a translation, I have chosen texts written in English, even though this has meant excluding a number of major works. Various themes in the scientific tradition can be followed through these selections. One is the relation between science and religion.[21] While the critics of the new philosophy accused it of promoting atheism, Bacon had already produced a reasoned defence of the investigation into God's created universe from the charge of prying into "forbidden knowledge", and the scientists represented here are orthodox Christians, like Bacon. Both Henry Power and Robert Hooke are moved by the discoveries revealed by the microscope to exclaim at the beauties of creation and the skills of God, while Boyle and Newton, like all their contemporaries, believed in God as the first cause of creation.

Another specifically scientific issue reflected here is the rise of the "mechanical" or "corpuscularian" philosophy, a movement that united several philosophical traditions, including Epicurean atomism and Aristotelian concepts of "minima".[22] In his *Sceptical Chymist* (1660) Robert Boyle, having destroyed both the Aristotelian and the Paracelsian theories of the elements, modestly and diffidently introduced his own belief that the ultimate constituents of the physical universe are matter and motion. In *The Excellency and Grounds of the Mechanical Hypothesis* (1674) he developed this theory, which by then

Hooke had also adopted.[23] The source for both men was Descartes, whose theory of matter had a considerable influence on English science, even though he was criticized for his tendency to create a scientific system by pure deduction rather than by observation and experiment. Here the Baconian influence is again evident, in the distrust of systems as forming a too premature conclusion to the enquiry (Bacon preferred to write in aphorisms to avoid the sclerosis imposed by a system; Boyle chose the essay for the same reason), and in the suspicion of *a priori* experimentally indefensible hypotheses. Baconian science was to be "grounded" upon experiment and observation, forming "solid" knowledge.

Science and language

Those typically Baconian metaphors for the new philosophy highlight one of the great critical issues in this period, the relation between science and language. A main stream of linguistic philosophy, from Aristotle to John Locke and beyond, had agreed that language was a conventional sign-system, where words represent concepts according to agreed social usage. As Bacon put it, "words are but the current tokens or marks of Popular Notions of things" (III, 338). Thinkers in this tradition were agreed in condemning the magical concept of language, in which words were not arbitrary combinations of letters and sounds but somehow contained the essence of the thing they denoted.[24] Yet while we can observe this broad distinction between the occult tradition and the new experimental philosophy, Bacon drew attention to some dangers in the way that non-occult scientists used language, dangers inherent in language itself. In the "Idols of the Market-place" he noted that the "ill and unfit choice of words" can produce confusion and "idle fancies". Two kinds of false appearance are created by words. The first occurs when *res* and *verba*, subject-matter and language, do not correspond. Words mislead when they "are either names of things which do not exist", "fantastic suppositions" such as "Fortune", the "Prime Mover" and the "Element of Fire", or such vague abstractions as the concept "humid", which cannot be "reduced to any constant meaning". The second illusion occurs when words are "names of things which exist, but yet confused and ill-defined, and hastily and irregularly derived from realities". Some thinkers have concluded that the problem of ambiguity can be overcome by the use of strict definitions, but Bacon saw that "even definitions cannot cure this evil in dealing with natural and material things; since the definitions themselves consist of words, and those words beget others: so that it is necessary to recur to individual instances" (IV, 61–2). Science must address the natural world directly while purging language of its imperfections.

In his *Parasceve*, or *Preparative towards a Natural and Experimental History*, appended to the *Novum Organum*, Bacon outlined the principles for collecting the observations of nature on which axioms or general laws would be established. The scientist must inspect nature at first hand, and do away "with antiquities, and citations of authors", that is, all literary traditions not tested against reality.

And for all that concerns ornaments of speech, similitudes, treasury of eloquence, and such like emptiness, let it be utterly dismissed. Also let all those
things which are admitted be themselves set down briefly and concisely, so
that they may be nothing less than words. For no man who is storing up
materials for ship-building or the like, thinks of arranging them elegantly . . .
all his care is that they be sound and good, and that they be so arranged as
to take up as little room as possible in the warehouse.   (IV, 254–5)

Bacon's injunctions apply only to the establishment of a body of scientific data, not to language as a whole. As Peter Shaw commented in 1733, "the business is not now to gain upon Men's Affections, or win them over to *Philosophy* by Eloquence, Similitudes, or the Art of Writing; which the author practised in *De Augmentis*; but carefully to enquire into, and firstly to copy and describe Nature as she is in herself; and there the Style cannot well be too plain and simple".[25] When the statutes of the Royal Society were published in 1728 they included the article that "In all Reports of experiments to be brought into the Society, the Matter of Fact shall be barely stated, without any Prefaces, Apologies, or Rhetorical Flourishes. . . ."[26]

Bacon's distinction between the proper language for recording observations and experiments, and language in all other contexts, where appeal from and to the imagination was legitimate, was correctly understood by many seventeenth-century writers. In his *History of the Royal Society* Sprat praised Bacon's "strong, cleer, and powerful Imaginations", expressed in a "vigorous and majestical" style, "the Wit Bold, and Familiar: The comparisons fetch'd out of the way, and yet the most easie" (p. 36). In their published work scientists showed themselves to be well aware of the hazards of language. Robert Boyle prefaced his collection of *Physiological Essays* (1661) with "Some Considerations touching Experimental Essays in general", preferring the essay for its unsystematic and discursive form, and announcing that as far as style is concerned, he had

9

endeavoured to write rather in a philosophical than a rhetorical strain, as desiring that my expressions should be rather clear and significant than curiously adorned ... And certainly in these discourses, where our design is only to inform readers, not to delight or persuade them, perspicuity ought to be esteemed at least one of the best qualifications of a style; and to affect needless rhetorical ornaments in setting down an experiment, or explicating something abstruse in nature, were little less improper than it were (for him that designs not to look directly upon the sun itself) to paint the eye-glasses of a telescope, whose clearness is their commendation, and in which even the most delightful colours cannot so much please the eye, as they would hinder the sight.[27]

Nevertheless, the "dull and insipid way of writing which is practised by many chymists" should be avoided, for a philosopher's style should not "disgust his reader by its flatness" (p. 304). Boyle's analogies are economical, and effective.

Robert Hooke was equally concerned to find the appropriate style for science. In his *General Scheme or Idea of the Present State of Natural Philosophy* he follows Bacon closely, cautioning that we be "very careful in what Sense we understand Philosophical Words already in use", which are bound to reflect prejudice and confusion. As things stand, the "Philosophical words of all languages ... seem to be for the most part very improper Marks set on confused and complicated Notions", so that "the Reason of a Man is very easily impos'd on by Discourse".[28] One remedy is to invent new words, then abandon the old and erroneous terms. In registering experiments the words chosen must contain nothing superfluous, and be

such as are shortest and express the Matter with the least Ambiguity and the greatest Plainness and Significancy ... avoiding all kinds of Rhetorical Flourishes, or Oratorical Garnishes ....

If possible the matter should be treated like "Geometrical Algebra, the expressing of many and very perplex Quantities by a few obvious and plain Symbols".[29] In the Cutlerian Lectures delivered to the Royal Society Hooke certainly followed his own recommendations. In the lectures on light, delivered in 1680, he complains that most extant definitions of light have confused the issue, writers having "spoken of it as it were Metaphorically and by Similitudes", their science consisting of "Rhetorical Embellishments, and no way tending to the Physical Explanation of its Effects and Proprieties" (that is, properties).[30] In his *Discourse of the Nature of Comets* (1682) Hooke pauses to take care

that I may not be mistaken in Expressions, and that the words I make use of, which are commonly used but by various Men are understood to signify various and very differing notions ....

He then explains the "Notions" he wishes to communicate through the terms "Body and Motion". Not untypically of seventeenth-century scientists he soon resorts to metaphor, describing them as "the *Male* and *Female* of Nature". Matter "is, as it were, the Female or Mother Principle", rightly called *"Materia,* or *Mater"*, being "without form, and void, and dark, a Power in itself wholly unactive, until it may be, as it were, impregnated by the second Principle, which may represent the *Pater*, and may be call'd *Paternus, Spiritus*, or hylarchick Spirit".[31] To show the interdependence of matter and motion Hooke has to use metaphor, careful though he is to qualify it with an "as it were".

The reform of language

In the traditional concept of language, words or *verba* expressed *res*, notions or subject-matter, with the primacy usually being given to content rather than form. Classical rhetoric-books included such advice as Cato's *Rem tene, verba sequentur* ("look after the subject-matter, and the words will look after themselves"), advice echoed by Cicero, Seneca and Horace.[32] Quintilian deplored this artificial separation: "Let care in words be solicitude for subject-matter. For generally the best words are inseparable from their subject-matter, and are discovered by their light."[33] The topic was revived during the Renaissance, often in the traditional form, as in Donne's remark that "language must wait upon matter, and words upon things". But, as A. C. Howell has shown, the concept of *res* gradually changed from "subject matter" to "things" in the sense of material objects.[34] Further, while the Aristotelian linguistic tradition, alive in Bacon, of defining words as arbitrary signs representing concepts, continued in Hobbes ("Names are signs not of things, but of our cogitations"), Locke (*"Words* ... come to be made use of by Men, as the Signs of their Ideas"), and other theorists of language,[35] another practice developed, especially in the polemic over language, which knocked out "concepts" or "ideas" and left "words" standing as the direct representations of "things". It became praiseworthy to discuss "so many things in so few words", or even to subordinate words to things.[36] In his *Ode to the Royal Society*, prefixed to Sprat's *History*, among the deeds for which that "mighty man" Francis Bacon is given such fulsome praise is that

> From Words, which are but Pictures of the Thought,
> (Though we our thoughts from them perversly drew)
> To things, the Mind's right Object, he it brought.[37]

Introduction

The dichotomy is too absolute, and too easy, but it does at least reflect an authentic theme in Bacon's philosophy.

Some discussions of this issue took the dichotomy to extremes. In his *History of the Royal Society*, commissioned by that body and closely supervised by John Wilkins, Thomas Sprat (then aged twenty-eight) included a description of the Society's "*manner of Discourse*" that has been much quoted. They have been "most sollicitous", Sprat alleges, about the language they have used,

> which, unless they had been very watchful to keep in due temper, the whole spirit and vigour of their *Design* had been soon eaten out by the luxury and redundance of *speech*. The ill effects of this superfluity of talking have already overwhelm'd most other *Arts* and *Professions*. (p. 111)

Sprat denounces the rhetorical "Ornaments of speaking" as degenerate, for they support the passions and subvert reason. He cannot "behold without indignation how many mists and uncertainties these specious *Tropes* and *Figures* have brought on our Knowledg", how "this vicious abundance of *Phrase*, this trick of *Metaphors*" has reached a stage almost beyond cure (p. 112). The Royal Society, however, has taken steps "towards the correcting of its excesses in *Natural Philosophy*", by enforcing

> a constant Resolution to reject all the amplifications, digressions and swellings of style: to return back to the primitive purity and shortness, when men deliver'd so many *things* almost in an equal number of *words*. They have exacted from all their members a close, naked, natural way of speaking; positive expressions; clear senses; a native easiness: bringing all things as near the Mathematical plainness as they can; and preferring the language of Artizans, Countrymen, and Merchants before that of Wits, or Scholars. (p. 113)

If any passage in Sprat's *History* shows up its intention to present the Royal Society's preferred image, this is it. Rational, experimental, coherent, balanced, clear, easy, solid, fruitful: all the approved epithets are applied to their style as much as to the content of their research. Yet the more we consider the implications of this account the less convincing it becomes. Traditionally it has been taken to prove that the Royal Society, or science, or puritanism, were hostile to rhetoric, or to language.[38] But the scholars who have argued this case have themselves included plenty of evidence that proves precisely the opposite.[39] Sprat has taken Bacon's account of the proper style for science and worked it up in an extravagant and almost hysterical way. Where is the evidence that most arts and professions have been "overwhelm'd" by "this superfluity of talking"? How could speech eat up "their Design"? Why is it that metaphors, traditionally the source of illumination and persuasion, are now seen as spreading "mists and uncertainties"? Are Sprat's own metaphors exempt from this charge? The one-to-one correlation of words and things would indeed recall a "primitive purity and shortness", but how could it serve to express complex ideas, such as the theories of Kepler or Galileo? Above all, where is the evidence that the members of the Royal Society have been constrained to practice "a close, naked" style, "as near the Mathematical plainness as they can"?

The fact is that many writers who disclaimed rhetoric were performing an old rhetorical ploy, a vanishing trick of eloquence to authenticate the speaker's sincerity, or to lull the listener or reader into the safe feeling that techniques of persuasion are not going to be used on him. Most such disclaimers are promptly followed by an effective use of rhetoric, working up the feelings, as in Sprat's own example of "indignation" and "disgust". Members of the Royal Society continued to use rhetoric, especially to attack their rivals: occultists, sectarians, and the opponents of science. In their polemics they claim to speak with a monopoly over reason, but the metaphors that some use are so violent that the reader alert to the tricks of rhetoric soon wonders whether they deceived themselves as well as hoping to deceive others. Sprat himself preserves rhetoric in its traditional role as the defender of good against evil (p. 111), assures students of grammar and rhetoric that the new "*Experimental Philosophy*" will not damage those arts (p. 324), and promises that science will actually offer a whole new world of similitudes and comparisons "solid and ... masculine" (pp. 414–15). Robert Boyle, his biographer informs us, mastered logic and rhetoric during his stay in Geneva,[40] and shows an impressive knowledge of rhetoric in his treatise *Some Considerations touching the Style of the Holy Scripture* (1661),[41] while his correspondents allude freely to such technical terms as *hyperbole*, *antiphrasis*, *periphrasis*, and *hysteron proteron*.[42]

As for Sprat's claim that scientists connected with the Royal Society abjured using metaphor, that is extraordinarily naive. No linguist or rhetorician would have dreamed of such a measure, since all (even Wilkins!) agreed that metaphor was fundamental to the extension of the vocabulary in any language, and to provide new shades of meaning. Many saw metaphor or other forms of analogy as essential to the discovery and communication of knowledge. Bacon stated that "there is no proceeding in invention of knowledge but by similitude" (III, 218). In transmitting knowledge analogy is also vital: "if any one wish to let new light on any subject into men's minds ...

he must still call in the aid of similitudes", which are "of prime use to the sciences, and sometimes indispensable" (IV, 698). Robert Hooke saw that the argument from analogy was essential in cases where distance or inaccessibility made direct observation impossible. In such cases

the best and utmost we can do ... is only accurately to observe all those Effects produced by them which fall within the Power of our Senses, and comparing them with like Effects produced by Causes that fall within the reach of our Senses, to examine and so from Sensibles to argue the Similitude of the nature of Causes that are wholly insensible.[43]

In his own writing Hooke explains the nature of light being generated in a luminous body by analogy with a stone mason striking a chisel with his hammer, which he hopes "may serve as a sensible [comprehensible] Similitude, by which I would inform the Understanding". It may "seem but a coarse Similitude for the Explication of the motion or action of Light, which is the most curious and spiritual of all sensible things", but "the more plain and obvious" the better it will serve to "inform the understanding of the manner how an operation which is too curious and fine to be reached by our Senses is performed". For if I can "comprehend and imagine one Local Motion that falls under the reach of my Senses, I can by similitude comprehend and understand another that is ten thousand Degrees below the reach of them".[44]

Hooke uses metaphor frequently, like every other writer of this period, both to explain or illuminate and to judge or evaluate, either in praise or dispraise. We have already seen his analogy from male and female to describe matter and motion; in the sections below Hooke talks of "the footsteps of Nature" which "are to be trac'd not only in her *ordinary course* but when she" makes "many *doublings* and *turnings*". The first paragraph of the first observation in *Micrographia* contains a long sustained analogy for the scientist following nature through her "meanders" and labyrinths looking for the necessary clue (p. 109, below). In his *Discourse of Earthquakes* (post 1668), discussing "the Defect of Natural History", namely that its descriptions of natural bodies are superficial and ambiguous, Hooke called for the setting up of a repository to contain

as full and compleat a Collection of all varieties of Natural Bodies as could be obtain'd, where an Inquirer might be able to have recourse, where he might peruse, and turn over, and spell, and read the Book of Nature, and observe the *Orthography, Etymologia, Syntaxis,* and *Prosodia* of Nature's Grammar, and by which, as with a *Dictionary,* he might readily turn to and find the true Figure, Composition, Derivation and Use of the Characters, Words, Phrases and Sentences of Nature written with indelible, and most exact, and most expressive letters, without which Books [one begins to wonder whether Hooke, pursuing his metaphor, has forgotten that he is describing a collection of natural bodies!] it will be very difficult to be a *Literatus* in the Language and Sense of Nature.   (*Posthumous Works*, p. 338)

That extended metaphor is worthy of Bacon, or Browne, and Hooke did not lose his ability to think in this way, constructing (in a lecture on 3 December 1690) an analogy, "The increase of Knowledge like the growth of a Plant", that covers a whole folio page (*ibid.*, p. 553). Robert Boyle was equally sensitive to metaphor, commenting on his own and other people's use of "metaphorical phrases" and "similes".[45] In his writing Boyle excelled in the evaluative use of metaphor, especially for polemic, as in the devastating analogies with which he attacks the alchemists' wilful obscurity in the opening paragraphs of *The Sceptical Chymist*, part IV: "they write darkly, and aenigmatically", and their readers are "stunned, as it were, or imposed upon by dark or empty words", and respond with "indignation at this unphilosophical way of teaching". Boyle's other enemy (following Bacon) was scholasticism, that mechanical application of Aristotelianism, and he describes the syllogisms of the scholastics as being "like the tricks of jugglers, whereby men doubt not but that they are cheated, though oftentimes they cannot declare by what slights they are imposed on". Sprat himself, of course, uses metaphor freely, to elevate the Royal Society and to confound its enemies. The "false Images" that Bacon identified "lye like Monsters" in the path of "*Modern Experimenters*". Their determination is such, however, that "if they cannot come at Nature in its particular *Streams*, they will have it in the *Fountain*". The alchemists publish their mysteries yet desire to conceal them, "which makes their *style* to resemble the *smoak* in which they deal". The mind of a man is "a Glass" or mirror, but "no *Magical Glass*, like that with which *Astrologers* use to deceive the Ignorant". The Baconian influence on these metaphors is clear, as in other images taken from building, harvests, fruit.

Henry Power's description of the minute universe perceived through his microscope makes so much use of metaphor and comparison that he seems unable to describe anything directly. The body of the fly is "studded with silver and black Armour", with "great black Bristles, like Porcupine quills"; "her wings look like a Sea-fan". The eye of the horsefly is "black and waved, or rather indented all over with a pure Emerauld-green, so that it looks like green silk Irish-stitch, drawn upon a black ground, and all latticed or chequered with dimples ... Her body looks like silver in frost-work, only fring'd all over with white silk". In his Conclusion Power produces many of those inspiring Baconian metaphors to encourage the lovers of natural philosophy: "You are the enlarged and Elastical souls of the world, who ... do make way for the Springy Intellect to flye out into its desired Expansion." This is "the Age wherein (me-thinks) Philosophy comes in with a Spring-tide ... Me-thinks I see how all the old Rubbish must be thrown away, and the rotten Buildings be overthrown and carried away with so powerful an Inundation."

The more we scrutinize Sprat's account of the Royal Society's supposedly naked and unrhetorical style the more fictitious it seems. He is to be seen not so much as a reformer who brought about a change in prose style but rather as a symptom of a general distrust of language in circles connected with the new philosophy. He had no effect on the style of Hooke, or Boyle, or Barrow, or Samuel Parker. The only evidence yet cited for a reform is a book by Joseph Glanvill, *The Vanity of Dogmatizing*, first published in 1661. When Glanvill reissued it in 1665 he gave it a new title, *Scepsis Scientifica*, adding a preface apologizing for its style, and declaring that he now preferred "a *natural* and *Unaffected Eloquence*" to "the *musick* and curiosity [affectation] of *fine Metaphors* and *dancing periods*". When he reused the material for a third time, in 1676, as the first essay in his *Essays on Several Important Subjects in Philosophy and Religion* he revised the style drastically, indeed almost too much so, for he wrote in the preface that although it is now "quite changed in the way of writing ... Methought I was somewhat fetter'd and tied down in doing it, and could not express myself with that ease, freedom and fullness which possibly I might have commanded amid fresh thoughts."[46] The explanation for the revisions may indeed be that Glanvill wished to "modernize" his style; but he was reducing a book to the length of an essay, and in any such merciless pruning the first things to go are the ornaments and graces. Yet Glanvill by no means abandoned metaphor, as any reader can see. I print specimen passages in the Appendix, also commenting on the scientific principle at issue.

Apart from this piece by Glanvill there is no other evidence that Sprat's account refers to any actual reform of prose-style, so that it must cease to be taken as a "manifesto" for reform but rather as a symptom of a general distrust of the abuse of language and rhetoric. One other symptom can be documented, however, in the writings of the man under whose tutelage Sprat wrote, and whose work he often echoed, John Wilkins. Wilkins expresses that phase of the disillusion with the deficiencies of language which took the form of inventing a new, artificial, "universal language", whether a phonetic or a sign-system, that could at one stroke correct all the "Idols of the Market-place". The proponents of universal language often claimed a relig-ious motivation, arguing that their scheme would repair the curse of tongues that had fallen on mankind at the Tower of Babel.[47] Wilkins had already advocated such a project in 1641, in his *Mercury; or the Secret and Swift Messenger*, citing Bacon's discussion of hieroglyphics and ciphers in the *Advancement of Learning*. There Bacon considered other modes of communication, not using the alphabet, such as those Eastern languages which employ "certain real characters, not nominal; characters, I mean, which represent neither letters nor words, but things and notions". Bacon still insists on "notions" as intermediate between words and things, and still uses the traditional linguistic concept of the conventional, agreed nature of the sign, such "real characters" being symbols "silently agreed on by custom" (IV, 439–40). (In fact, of course, Chinese uses ideograms, in which the symbol is at times a picture of the thing.) Bacon criticized these systems since they must need a "vast multitude" of signs, as many signs as "there are radical words", and Wilkins may have been challenged by that comment to devise a more efficient system when he produced his *Essay Towards a Real Character and a Philosophical Language* (1668): "philosophical" here means "scientific", as so often at this time.

In his *Epistle* Wilkins states that "as *things* are better than *words*, as *real knowledge* is beyond *elegancy of speech*" (Sig. a₂ʳ) his system will not only provide the best way to attain "real knowledge" but will also reduce controversies in religion and politics by eliminating error and confusion (b₁ʳ). The task is urgent, given the proliferation of ambigu-ity "in late times", where a "grand imposture of Phrases hath almost eaten out solid knowledge in all professions" (p. 18). Wilkins pro-poses to transcend the normal deficiencies of language, first by devising a "distinct *Mark* to represent every thing"; secondly, by arranging the names of things so as "to contain such a kind of *affinity* or *opposition* in their letters and sounds, as might be some way answerable to the nature of the things which they signified" (p. 21). He sees himself bound to make "the difficult attempt of enumerating and describing all such things and notions as fall under discourse" (p. 22). That is, Wilkins will first classify the world, categorize the whole of objective reality into a logical sequence, genus and species forming a detailed taxonomy.[48] Then, on the facing page, as it were, he will construct, first, a series of signs to represent each category in a written language, and then produce a synthetic arrangement of vowels and consonants to give a spoken form.

It is a grand scheme, but hopelessly impracticable from the outset. How can we classify reality, once and for all? Even if it could be done once, to do so would either deny all further scientific progress, so defeating the growth of "real knowledge", or it would mean endless revision of the sign-systems. In fact, the deficiencies of Wilkins's scheme became evident while he was working on it, for he employed a colleague in the Royal Society, the outstanding botanist John Ray, to help him draw up tables of plants and organic substances. But, Ray complained, there were simply more of these than could be accommodated in Wilkins's master plan, which had to be arranged into three main groups and nine subordinate ones to fulfil a trinitarian theory.[49] As for the sign-system, with its combination of straight, angular, and curved lines, that soon proved insufficient. Thus ⌐ has to represent "How", "More", and "Less", while ⌐ stands for "So", "Most", and "Least" (pp. 385ff.). If the same sign has to mean both larger and smaller, then the potential for ambiguity is far greater than anything in language as we know it. Our language, which is symbolic, not "real", natural, not "philosophical" or scientific, with its limited number of symbols and sounds, and the all-important principle of the freedom of the sign to represent any idea or object, is able to represent an infinite and ever-growing world. Paradoxically, perhaps, for someone so committed to the Baconian vision of continuous scientific advancement, Wilkins would have frozen language into a finite, and above all impracticable, system.

Latin and science

Although most of Thomas Sprat's account of the Royal Society's reform of prose can be dismissed as mythical, it does reflect some contemporary crises. One of these is alluded to in his account of the "native easiness" they wished for, the language of "Artizans ... and Merchants": that is, English, rather than that of "Wits, or Scholars". "Scholars" could read, write, and speak Latin, since it was the major and sometimes the only language of instruction in schools and universities. Latin was also the medium for the international scholarly community, and English scientists continued to write major works in Latin: Bacon, William Harvey, John Ray, Isaac Newton. Much of the extant scientific literature was in Latin, so that even those scientists who preferred to write in English – Boyle, Hooke, Charleton, among others – had to find English equivalents for technical terms if they wished to advance learning in the vernacular. Although the high point of neologizing, importing words from other languages, has been placed in the Elizabethan period, a glance at the *Chronological English Dictionary* will confirm the impression gained from ordinary reading, that a vast number of Latinate words continued to be coined throughout the seventeenth century.[50] Sir Thomas Browne commented in 1646 that "if elegancy still proceedeth, and English Pens maintain that stream we have of late observed to flow from many, we shall within few years be fain to learn Latine to understand English, and a work will prove of equal facility in either".[51] Browne wrote in the vernacular, although he feared that the complexity of his subject-matter would "sometimes carry us into expressions beyond meer English apprehensions". In 1650 Walter Charleton, a chemist and translator, praised the "thousands of forraigne Words providently brought home" from the classical and European languages in recent years, which have become "so familiar to us that now even Children speak much of Latin before they can well read a word of English". As for the effect this mingling of tongues had on Charleton's prose-style, a recent commentator has described his *Physiologia Epicuro-Gassendo-Charletoniana* (1654: in English, despite its title) as "a hard read: the English is so dreadfully Latinized that it is almost difficult to tell, at some points, in which language he believed himself to be thinking".[52]

This state of affairs aroused concern in some quarters. In 1664 the Royal Society set up a committee "for improving the English tongue", doubtless imitating the Académie Française. John Evelyn, unable to attend a meeting in 1665, sent an account of the many changes suffered by English in his time, with some proposed remedies. A lexicon of "all the pure English words" should be compiled – a lost cause, surely, even then – so that "no innovation might be us'd or favour'd" until necessary. Another lexicon, of established "technical words", should be prepared, and thirdly "a full catalogue of exotic words, such as are daily minted by our *Logodaedali*" (word-inventors), for this lust for new coinages "will in time quite disguise the language" unless restrained.[53] With our greater understanding of linguistic change we can see the futility of trying to shackle a language. Scientists of the day defended themselves on the grounds that no equivalent existed in English for the "hard words" or technical terms of the sciences. Sir Kenelm Digby complained of "the scarcity of our language" not affording him "apt words of our own to expresse significantly" his notions, so that he has had "to borrow them from the Latine Schoole, where there is much adoe about them".[54] Boyle summarized the arguments of the alchemist Sennertus in the original Latin, so "that I might also retaine the propriety of some Latine termes, to which I do not readily remember any that fully answer in English".[55]

Justifiable, even unavoidable though this neologizing may have been, the effect was to create a strangely hybrid tongue. One would be ill-advised to attempt to read extensively in seventeenth-century scientific English without the *OED* or a good Latin–English dictionary at one's elbow. In Boyle, who writes on the whole a clear, straightforward, if prolix and verbose English, in *The Sceptical Chymist* alone we find such technical terms as "dissipable", "comminution", "factitious concrete", "colliquated", "candent", "calcination", "dulcified colcothar", "compurgator", "idoneous", "ebullition", "affused", "cohobations", "diaphoretic and very deopilative", "opacous", "feculancy", and many more. The problem of an inadequate English vocabulary for science was already an issue a century earlier, and scientific works in English included glossaries of new words.[56] When Richard Tomlinson published a translation of Renodaeus' *Dispensatory* (1657) the publisher felt obliged to give its purchasers *A Physical Dictionary. Or, an Interpretation of such crabbed Words and Terms of Art as are derived from the Greek or Latin*, an aid useful to all, "especially such as are not Scholars". Tomlinson was a minor and eccentric figure, but an equally deterrent list of Latinisms could be derived from the writings of Robert Hooke, or any other scientist using English. Indeed, when Thomas Shadwell satirized the new philosophy in *The Virtuoso* (1676), Sir Nicholas Gimcrack, the untiring and indiscriminate advocate of experiment, is a cruel caricature of Hooke himself, as the unfortunate butt realized when he saw the play.[57] The language Shadwell gives his dummy is in effect a tissue of words used by Hooke, among others: "diffusive", "precelling in physico-mechanical investigations", "follicular impulsion", "testaceous", "stentrophonical", "docible" and the like. The most extended parody is the description of how a plum turns blue: "It comes first to fluidity, then to orbiculation, then fixation, so to angulization, then crystallization, from thence to germination or ebullition, then vegetation, then plantamination, perfect animation, sensation, local motion, and the like."[58] The model for this flight of fancy is none other then Hooke's *Micrographia* itself, in Observation xx (p. 124 below). This much credence, then, we should give to Sprat's account: on the issue of Latinisms the Royal Society had grounds to put its house in order. If only the English language could have permitted it!

Conclusion

Sprat's diagnosis of the reforms needed (setting aside whether they were ever made) can now be seen partly as fair comment on the state of English prose, partly as exaggerated and over-dramatic. If it loses status as an account of what actually happened in London in the 1660s, it can still be taken as a prophetic document – that is, what Sprat falsely announced as a *fait accompli* then is what scientific prose has become: "Close, naked", unadorned by rhetorical ornament; preferring "positive expressions; clear senses; ... Mathematical plainness". This is the style in which our colleagues write their research papers and their textbooks, a style which, it seems safe to predict, will not much engage future historians of English prose. Yet seventeenth-century writers, happily enough, had not reached this "degree zero" of writing, this absence of style. In their work we still find a fluent use of metaphor, a fresh, at times child-like excitement at the "new visible World" revealed by the microscope, an eagerness to understand nature and restore the power over her that man lost at the Fall. They record laboratory experiments, but with a vividness and immediacy that brings us into direct contact with the whole event. Robert Boyle and his assistants extract air from their cylinder, causing the lamb's bladder inside to swell up and burst, "with so loud and vehement a noise as stunned those that were by, and made us for a while after almost deaf" (p. 58, below). Boyle also performs experiments to establish how long animals can live without air, finding that a lark deprived of air for three quarters of an hour fell into "violent and irregular convulsions", which he tries to prevent by letting in more air: but "the whole tragedy had been concluded" (p. 64, below). Happily, in another experiment Boyle leaves a hungry mouse in his air-pump over night, puts the cylinder "by the fire-side, to keep him from being destroyed by the immoderate cold of the frosty night", and finds next morning that the mouse is alive and well, having eaten most of "the cheese that had been put in with him" (p. 66, below). Robert Hooke's difficulties with the insects he tried to fix under his microscope come to a climax with the ant, who will not lie still. If "its feet were fetter'd in Wax or Glue it would so twist and wind its body" that Hooke could not get a good view of it; if he killed it, its body was then damaged, or would shrivel up. So he finally takes a large specimen and gives it "a Gill of Brandy or Spirit of Wine, which after a while e'en knock'd him down dead drunk, so that he became moveless". But as the narcotic wears off, the ant, as if woken "out of a drunken sleep ... suddenly reviv'd and ran away" (p. 131, below). The spectacle of Hooke chasing an ant around his laboratory and giving it "knock-out drops" so that he can enlarge it and draw it, gives a liveliness to the whole proceeding that we can feel to this day.

In their contact with the natural world, too, these writers manage to convey their activities freshly. Hooke prods around cliffs on the south coast and the Isle of Wight, digging out shells and stones, some of them "so very soft that I could easily with my foot crush them and make Impressions into them, and could thrust a Walking-stick I had in my Hand a great depth into them" (p. 148, below). Newton, setting up his famous experiment with the prism, closes the shutters (either in his room in Great Court, Trinity, or at his house at Woolsthorpe[59]) and then makes "a small hole" in them, "whose diameter may conveniently be about a third part of an inch" (p. 208, below). The room must offer a depth of twenty-two feet from the window (either room would have been big enough), but just how he discovered that is not revealed in this account of the ideal experiment, where each stage leads smoothly to the next. Immediacy, precision, rational organization – these are some of the qualities that characterize the scientific prose of this period. True, the Latinate jargon remains an obstacle that the English language has overcome, shedding many of its coinages, domesticating others. The desire to reduce language to mathematical signs failed, for obvious reasons, as did the attempt to outlaw the expressive devices of rhetoric. All scientists need the imagination to understand nature, but whereas today that power is expressed mathematically or physically, in seventeenth-century science imagination still invested language itself.

Francis Bacon

(1561–1626)

1 Preparative towards a Natural and Experimental History (1620)

The publication in 1620 of the *Instauratio Magna*, the 'Great Instauration' (restoration or renewal), marked Bacon's most ambitious attempt to re-direct the course of science. The whole scheme was to consist of six parts, of which he only completed fragments. The *Novum Organum* or 'New Instrument' of scientific method, with two of the projected eight books completed, was the most substantial work of theory that he issued. In the five years of life remaining, apart from publishing some major literary works (the *Essays*, the *History of Henry VII*), Bacon devoted most of his scientific labours to preparing specimen parts of the 'Natural and Experimental History', the prime collection of scientific data, that for him formed the first step in all scientific research. Yet he also stressed that material could not be collected, nor experiments performed, without a controlling theory and plan.

Source: *Parasceve ad Historiam Naturalem et Experimentalem*, appended to the *Novum Organum* in the *Instauratio Magna* (1620). Translation by James Spedding in *The Works of Francis Bacon*, ed. J. Spedding, R. L. Ellis, and D. D. Heath, 14 vols. (London, 1857–1874), vol. 4, pp. 249–71 (Latin text: vol. 1, pp. 391–411).

Description of a NATURAL and EXPERIMENTAL HISTORY, such as may serve for the foundation of a true philosophy

My object in publishing my Instauration by parts is that some portion of it may be put out of peril.[1] A similar reason induces me to subjoin here another small portion of the work, and to publish it along with that which has just been set forth. This is the description and delineation of a Natural and Experimental History such as may serve to build philosophy upon, and containing material true and copious and aptly digested for the work of the Interpreter which follows. The proper place for it would be when I come in due course to the *Preparatives* of Inquiry. I have thought it better however to introduce it at once without waiting for that. For a history of this kind, such as I conceive and shall presently describe, is a thing of very great size, and

23

cannot be executed without great labour and expense; requiring as it does many people to help, and being (as I have said elsewhere) a kind of royal work. It occurs to me therefore that it may not be amiss to try if there be any others who will take these matters in hand; so that while I go on with the completion of my original design, this part which is so manifold and laborious may even during my life (if it so please the Divine Majesty) be prepared and set forth, others applying themselves diligently to it along with me; the rather because my own strength (if I should have no one to help me) is hardly equal to such a province. For as much as relates to the work itself of the intellect, I shall perhaps be able to master that by myself; but the materials on which the intellect has to work are so widely spread that one must employ factors and merchants to go everywhere in search of them and bring them in. Besides I hold it to be somewhat beneath the dignity of an undertaking like mine that I should spend my own time in a matter which is open to almost every man's industry. That however which is the main part of the matter I will myself now supply, by diligently and exactly setting forth the method and description of a history of this kind, such as shall satisfy my intention; lest men for want of warning set to work the wrong way, and guide themselves by the example of the natural histories now in use, and so go far astray from my design. Meanwhile what I have often said I must here emphatically repeat; that if all the wits of all the ages had met or shall hereafter meet together; if the whole human race had applied or shall hereafter apply themselves to philosophy, and the whole earth had been or shall be nothing but academies and colleges and schools of learned men; still without a natural and experimental history such as I am going to prescribe, no progress worthy of the human race could have been made or can be made in philosophy and the sciences. Whereas on the other hand, let such a history be once provided and well set forth, and let there be added to it such auxiliary and light-giving experiments as in the very course of interpretation will present themselves or will have to be found out; and the investigation of nature and of all sciences will be the work of a few years. This therefore must be done, or the business must be given up. For in this way, and in this way only, can the foundations of a true and active philosophy be established; and then will men wake as from deep sleep, and at once perceive what a difference there is between the dogmas and figments of the wit and a true and active philosophy, and what it is in questions of nature to consult nature herself.

First then I will give general precepts for the composition of this history; then I will set out the particular figure of it, inserting sometimes as well the purpose to which the inquiry is to be adapted and referred as the particular point to be inquired; in order that a good understanding and forecast of the mark aimed at may suggest to men's minds other things also which I may perhaps have overlooked. This history I call *Primary History*, or the *Mother History*.

APHORISMS on the COMPOSITION OF THE PRIMARY HISTORY

APHORISM

I.

Nature exists in three states, and is subject as it were to three kinds of regimen. Either she is free, and develops herself in her own ordinary course; or she is forced out of her proper state by the perverseness and insubordination of matter and the violence of impediments; or she is constrained and moulded by art and human ministry. The first state refers to the *species* of things; the second to *monsters*; the third to *things artificial*. For in things artificial nature takes orders from man, and works under his authority: without man, such things would never have been made. But by the help and ministry of man a new face of bodies, another universe or theatre of things, comes into view. Natural History therefore is threefold. It treats of the *liberty* of nature, or the *errors* of nature, or the *bonds* of nature: so that we may fairly distribute it into history of *Generations*, of *Pretergenerations*,* and of *Arts*; which last I also call *Mechanical* or *Experimental* history. And yet I do not make it a rule that these three should be kept apart and separately treated. For why should not the history of the monsters in the several species be joined with the history of the species themselves? And things artificial again may sometimes be rightly joined with the species, though sometimes they will be better kept separate. It will be best therefore to consider these things as the case arises. For too much method produces iterations and prolixity as well as none at all.

II.

Natural History, which in its subject (as I said) is threefold, is in its use twofold. For it is used either for the sake of the knowledge of the particular things which it contains, or as the primary material of philosophy and the stuff and subject-matter of true induction. And it is this latter which is now in hand; now, I say, for the first time: nor has it ever been taken in hand till now. For neither Aristotle, nor Theophrastus, nor Dioscorides, nor Caius Plinius, ever set this before them as the end of natural history. And the chief part of the matter rests in this: that they who shall hereafter take it upon them to write natural history should bear this continually in mind – that they ought not to consult the pleasure of the reader, no nor even that utility which may be derived immediately from their narrations; but to seek out and gather together such store and variety of things as may suffice for the formation of true axioms. Let them but remember this, and they will find out for themselves the method in which the history should be composed. For the end rules the method.

III.

But the more difficult and laborious the work is, the more ought it to be discharged of matters superfluous. And therefore there are three things upon which men should be warned to be sparing of their labour – as those which will immensely increase the mass of the work, and add little or nothing to its worth.

First then, away with antiquities, and citations or testimonies of authors; also with disputes and controversies and differing opinions; everything in short which is philological. Never cite an author except in a matter of doubtful credit: never introduce a controversy unless in a matter of great moment. And for all that concerns ornaments of speech, similitudes, treasury of eloquence, and such like emptinesses, let it be utterly dismissed. Also let all those things which are admitted be themselves set down briefly and concisely, so that they may be nothing less than words. For no man who is collecting and storing up materials for ship-building or the like, thinks of arranging them elegantly, as in a shop, and displaying them so as to please the eye; all his care is that they be sound and good, and that they be so arranged as to take up as little room as possible in the warehouse. And this is exactly what should be done here.

Secondly, that superfluity of natural histories in descriptions and pictures of species, and the curious variety of the same, is not much to the purpose. For small varieties of this kind are only a kind of sports and wanton freaks of nature; and come near to the nature of individuals. They afford a pleasant recreation in wandering among them and looking at them as objects in themselves; but the information they yield to the sciences is slight and almost superfluous.

Thirdly, all superstitious stories (I do not say stories of prodigies, when the report appears to be faithful and probable; but superstitious stories) and experiments of ceremonial magic should be altogether rejected. For I would not have the infancy of philosophy, to which natural history is as a nursing-mother, accustomed to old wives' fables. The time will perhaps come (after we have gone somewhat deeper into the investigation of nature) for a light review of things of this kind; that if there remain any grains of natural virtue in these dregs, they may be extracted and laid up for use. In the meantime they should be set aside. Even the experiments of natural magic should be sifted diligently and severely before they are received; especially those which are commonly derived from vulgar sympathies and antipathies, with great sloth and facility both of believing and inventing.

And it is no small thing to relieve natural history from the three superfluities above mentioned, which would otherwise fill volumes. Nor is this all. For in a great work it is no less necessary that what is admitted should be written succinctly than that what is superfluous should be rejected; though no doubt this kind of chastity and brevity will give less pleasure both to the reader and the writer. But it is always to be remembered that this which we are now about is only a granary and storehouse of matters, not meant to be pleasant to stay or live in, but only to be entered as occasion requires, when anything is wanted for the work of the *Interpreter*, which follows.

IV.

In the history which I require and design, special care is to be taken that it be of wide range and made to the measure of the universe. For the world is not to be narrowed till it will go into the understanding (which has been done hitherto), but the understanding to be expanded and opened till it can take in the image of the world, as it is in fact. For that fashion of taking few things into account, and pronouncing with reference to a few things, has been the ruin of everything. To resume then the division of natural history which I made just now, – viz. that it is a history of Generations, Pretergenerations, and Arts, – I divide the History of Generations into five parts. The first, of Ether and things Celestial. The second, of Meteors and the regions (as they call them) of Air; viz. of the tracts which lie between the moon and the surface of the earth; to which part also (for order's sake, however the truth of the thing may be) I assign Comets of whatever kind, both higher and lower. The third, of Earth and Sea. The fourth, of the Elements (as they call them), flame or fire, air, water, earth: understanding however by Elements, not the first principles of things, but the greater masses of natural bodies. For the nature of things is so distributed that the quantity or mass of some bodies in the universe is very great, because their configurations require a texture of matter easy and obvious; such as are those four bodies which I have mentioned; while of certain other bodies the quantity is small and weakly supplied, because the texture of matter which they require is very complex and subtle, and for the most part determinate and organic; such as are the species of natural things, – metals, plants, animals. Hence I call the former kind of bodies the *Greater Colleges*, the latter the *Lesser Colleges*. Now the fourth part of the history is of those Greater Colleges – under the name of Elements, as I said. And let it not be thought that I confound this fourth part with the second and third, because in each of them I have mentioned air, water, and earth. For the history of these enters into the second and third, as they are integral parts of the world, and as they relate to the fabric and configuration of the universe. But in the fourth is contained the history of their own substance and nature, as it exists in their several parts of uniform structure, and without reference to the whole. Lastly, the fifth part of the history contains the Lesser Colleges, or Species; upon which natural history has hitherto been principally employed.

As for the history of Pretergenerations, I have already said that it may be most conveniently joined with the history of Generations; I mean the history of prodigies which are natural. For the superstitious history of marvels (of whatever kind) I remit to a quite separate treatise of its own; which treatise I do not wish to be undertaken now at first, but a little after, when the investigation of nature has been carried deeper.

History of Arts, and of Nature as changed and altered by Man, or Experimental History, I divide into three. For it is drawn either from mechanical arts, or from the operative part of the liberal arts; or from a number of crafts and experiments which have not yet grown into an art properly so called, and which sometimes indeed turn up in the course of most ordinary experience, and do not stand at all in need of art.

As soon therefore as a history has been completed of all these things which I have mentioned, namely, Generations, Pretergenerations, Arts and Experiments, it seems that nothing will remain unprovided whereby the sense can be equipped for the information of the understanding. And then shall we be no longer kept dancing within little rings, like persons bewitched, but our range and circuit will be as wide as the compass of the world.

v.

Among the parts of history which I have mentioned, the history of Arts is of most use, because it exhibits things in motion, and leads more directly to practice. Moreover it takes off the mask and veil from natural objects, which are commonly concealed and obscured under the variety of shapes and external appearance. Finally, the vexations of art are certainly as the bounds and handcuffs of Proteus,[2] which betray the ultimate struggles and efforts of matter. For bodies will not be destroyed or annihilated; rather than that they will turn themselves into various forms. Upon this history therefore, mechanical and illiberal* as it may seem, (all fineness and daintiness set aside) the greatest diligence must be bestowed.

Again, among the particular arts those are to be preferred which exhibit, alter, and prepare natural bodies and materials of things; such as agriculture, cookery, chemistry, dyeing; the manufacture of glass, enamel, sugar, gunpowder, artificial fires, paper, and the like. Those which consist principally in the subtle motion of the hands or instruments are of less use; such as weaving, carpentry, architecture, manufacture of mills, clocks, and the like; although these too are by no means to be neglected, both because many things occur in them which relate to the alterations of natural bodies, and because they give accurate information concerning local motion, which is a thing of great importance in very many respects.

But in the whole collection of this history of Arts, it is especially to be observed and constantly borne in mind, that not only those experiments in each art which serve the purpose of the art itself are to be received, but likewise those which turn up anyhow by the way. For example, that locusts or crabs, which were before of the colour of mud, turn red when baked, is nothing to the table; but this very instance is not a bad one for investigating the nature of redness, seeing that the same thing happens in baked bricks. In like manner the fact that meat is sooner salted in winter than in summer is not only important for the cook that he may know how to regulate the pickling, but is likewise a good instance for showing the nature and impression of cold. Therefore it would be an utter mistake to suppose that my intention would be satisfied by a collection of experiments of arts made only with the view of thereby bringing the several arts to greater perfection. For though this be an object which in many cases I do not despise, yet my meaning plainly is that all mechanical experiments should be as streams flowing from all sides into the sea of philosophy. But how to select the more important instances in every kind (which are principally and with the greatest diligence to be sought and as it were hunted out) is a point to be learned from the prerogatives of instances.

VI.

In this place also is to be resumed that which in the 99th, 119th, and 120th Aphorisms of the first book[3] I treated more at large, but which it may be enough here to enjoin shortly by way of precept; namely, that there are to be received into this history, first, things the most ordinary, such as it might be thought superfluous to record in writing, because they are so familiarly known; secondly, things mean, illiberal, filthy (for "to the pure all things are pure," and if money obtained from Vespasian's tax smelt well,[4] much more does light and information from whatever source derived); thirdly, things trifling and childish (and no wonder, for we are to become again as little children); and lastly, things which seem over subtle, because they are in themselves of no use. For the things which will be set forth in this history are not collected (as I have already said) on their own account; and therefore neither is their importance to be measured by what they are worth in themselves, but according to their indirect bearing upon other things, and the influence they may have upon philosophy.

VII.

Another precept is, that everything relating both to bodies and virtues in nature be set forth (as far as may be) numbered, weighed, measured, defined. For it is works we are in pursuit of, not speculations; and practical working comes of the due combination of physics and mathematics. And therefore the exact revolutions and distances of the planets – in the history of the heavenly bodies; the compass of the land and the superficial space it occupies in comparison of the waters – in the history of earth and sea; how much compression air will bear without strong resistance – in the history of air; how much one metal outweighs another – in the history of metals; and numberless other particulars of that kind are to be ascertained and set down. And when exact proportions cannot be obtained, then we must have recourse to indefinite estimates and comparatives. As for instance (if we happen to distrust the calculations of astronomers as to the distances of the planets), that the moon is within the shadow of the earth; that Mercury is beyond the moon; and the like. Also when mean proportions cannot be had, let extremes be proposed; as that a weak magnet will raise so many times its own weight of iron, while the most powerful will raise sixty times its own weight (as I have myself seen in the case of a very small armed* magnet). I know well enough that these definite instances do not occur readily or often, but that they must be sought for as auxiliaries in the course of interpretation itself when they are most wanted. But nevertheless if they present themselves accidentally, provided they do not too much interrupt the progress of the natural history, they should also be entered therein.

VIII.

With regard to the credit of the things which are to be admitted into the history; they must needs be either certainly true, doubtful whether true or not, or certainly not true. Things of the first kind should be set down simply; things of the second kind with a qualifying note, such as "it is reported", "they relate", "I have heard from a person of credit", and the like. For to add the arguments on either side would be too laborious and would certainly interrupt the writer too much. Nor is it of much consequence to the business in hand; because (as I have said in the 118th Aphorism of the first book) mistakes in experimenting, unless they abound everywhere, will be presently detected and corrected by the truth of axioms.[5] And yet if the instance be of importance, either from its own use or because many other things may depend upon it, then certainly the name of the author should be given; and not the name merely, but it should be mentioned withal whether he took it from report, oral or written (as most of Pliny's statements are), or rather affirmed it of his own knowledge; also, whether it was a thing which happened in his own time or earlier; and again whether it was a thing of which, if it really happened, there must needs have been many witnesses; and finally whether the author was a vain-speaking and light person, or sober and severe; and the like points, which bear upon the weight of the evidence. Lastly things which though certainly not true are yet current and much in men's mouths, having either through neglect or from the use of them in similitudes prevailed now for many ages, (as that the diamond binds the magnet, garlic weakens it; that amber attracts everything except basil; and other things of that kind) these it will not be enough to reject silently; they must be in express words proscribed, that the sciences may be no more troubled with them.

Besides, it will not be amiss, when the source of any vanity or credulity happens to present itself, to make a note of it; as for example, that the power of exciting Venus is ascribed to the herb *Satyrion*, because its root takes the shape of testicles; when the real cause of this is that a fresh bulbous root grows upon it every year, last year's root still remaining; whence those twin bulbs. And it is manifest that this is so; because the new root is always found to be solid and succulent, the old withered and spongy. And therefore it is no marvel if one sinks in water and the other swims; which nevertheless goes for a wonder, and has added credit to the other virtues ascribed to this herb.

IX.

There are also some things which may be usefully added to the natural history, and which will make it fitter and more convenient for the work of the interpreter, which follows. They are five.

First, questions (I do not mean as to causes but as to the fact) should be added, in order to provoke and stimulate further inquiry; as in the history of Earth and Sea, whether the Caspian ebbs and flows, and at how many hours' interval; whether there is any Southern Continent, or only islands; and the like.

Secondly, in any new and more subtle experiment the manner in which the experiment was conducted should be added, that men may be free to judge for themselves whether the information obtained from that experiment be trustworthy or fallacious; and also that men's industry may be roused to discover if possible methods more exact.

Thirdly, if in any statement there be anything doubtful or questionable, I would by no means have it suppressed or passed in silence, but plainly and perspicuously set down by way of note or admonition. For I want this primary history to be compiled with a most religious care, as if every particular were stated upon oath; seeing that it is the book of God's works, and (so far as the majesty of heavenly may be compared with the humbleness of earthly things) a kind of second Scripture.

Fourthly, it would not be amiss to intersperse observations occasionally, as Pliny has done; as in the history of Earth and Sea, that the figure of the earth (as far as it is yet known) compared with the seas, is narrow and pointed towards the south, wide and broad towards the north; the figure of the sea contrary: – that the great oceans intersect the earth in channels running north and south, not east and west; except perhaps in the extreme polar regions. It is also very good to add canons (which are nothing more than certain general and catholic observations); as in the history of the Heavenly Bodies, that Venus is never distant more than 46 parts from the sun; Mercury never more than 23; and that the planets which are placed above the sun move slowest when they are furthest from the earth, those under the sun fastest. Moreover there is another kind of observation to be employed, which has not yet come into use, though it be of no small importance. This is, that to the enumeration of things which are should be subjoined an enumeration of things which are not. As in the history of the Heavenly Bodies, that there is not found any star oblong or triangular, but that every star is globular; either globular simply, as the moon; or apparently angular, but globular in the middle, as the other stars; or apparently radiant but globuar in the middle, as the sun; – or that the stars are scattered about the sky in no order at all; so that there is not found among them either quincunx or square, or any other regular figure (howsoever the names be given of Delta, Crown, Cross, Chariot, &c.) – scarcely so much as a straight line; except perhaps in the belt and dagger of Orion.

Fifthly, that may perhaps be of some assistance to an inquirer which is the ruin and destruction of a believer; viz. a brief review, as in passage, of the opinions now received, with their varieties and sects; that they may touch and rouse the intellect, and no more.

x.

And this will be enough in the way of general precepts; which if they be diligently observed, the work of the history will at once go straight towards its object and be prevented from increasing beyond bounds. But if even as here circumscribed and limited it should appear to some poor-spirited person a vast work – let him turn to the libraries; and there among other things let him look at the bodies of civil and canonical law on one side, and at the commentaries of doctors and lawyers on the other; and see what a difference there is between the two in point of mass and volume. For we (who as faithful secretaries do but enter and set down the laws themselves of nature and nothing else) are content with brevity, and almost compelled to it by the condition of things; whereas opinions, doctrines, and speculations are without number and without end.

And whereas in the Plan of the Work I have spoken of the *Cardinal Virtues* in nature, and said that a history of these must also be collected and written before we come to the work of Interpretation;[6] I have not forgotten this, but I reserve this part for myself; since until men have begun to be somewhat more closely intimate with nature, I cannot venture to rely very much on other people's industry in that matter.

And now should come the delineation of the particular histories. But I have at present so many other things to do that I can only find time to subjoin a Catalogue of their titles. As soon however as I have leisure for it, I mean to draw up a set of questions on the several subjects, and to explain what points with regard to each of the histories are especially to be inquired and collected, as conducing to the end I have in view, – like a kind of particular Topics.* In other words, I mean (according to the practice in civil causes) in this great

Francis Bacon (1561–1626)

Plea or Suit granted by the divine favour and providence (whereby the human race seeks to recover its right over nature), to examine nature herself and the arts upon interrogatories.*

[Bacon appends a 'Catalogue of Particular Histories by Titles', itemizing 130 separate topics of investigation into astronomy, meteorology, geology, botany, anatomy, crafts, agriculture, and much else.]

2 New Atlantis (c.1624)

Bacon's philosophical fable alludes to Plato's account (in *Timaeus* 24–5 and *Critias* 108–11) of the island Atlantis, off the straits of Gibraltar, that had sunk into the sea some 9,000 years before Solon. The work begins like a typical Elizabethan voyage narrative, recording how some travellers, blown off their route by strong West winds, are becalmed on the unknown island of Bensalem. They are kindly received by the inhabitants, taken to the Strangers' House, well fed and nursed. The Governor tells them of the island's institution of "Salomons House, which House or Colledge (my good Brethren) is the very Eye of this Kingdome". Bacon then uses this framework to project an ideal scientific research institute, embracing the whole sequence of investigation, from empirical observation and experiment to the analysis of current research, the formation of theories, and their redirection for further research. Some of his projections were surprisingly far-reaching.

Source: *New Atlantis. A Worke Unfinished*, appended to *Sylva Sylvarum: or A Naturall Historie* (London, 1624), ed. W. Rawley. In his preface Rawley (Bacon's chaplain) explained that "This Fable my Lord devised to the end that He might exhibit therein a Modell or Description of a Colledge[7] instituted for the Interpreting of Nature and the Producing of Great and Marveilous Works for the Benefit of Men; Under the name of Salomon's House, or the Colledge of the Six Dayes' Works. And even so farre his Lordship hath proceeded, as to finish that Part. Certainly, the Modell is more Vast and High than can possibly be imitated in all things; Notwithstanding most Things therein are within Mens Power to effect." The *New Atlantis* was reprinted with the *Sylva Sylvarum* 13 times between 1627 and 1685; it was issued separately in 1659, in French translation (1631, 1652 with a continuation by G. B. Raguet), and in Latin (1648, 1661). An English continuation by "R. H. Esquire" (probably Robert Hooke) appeared in 1660.

[The Governor of the island informs the travellers of the rule of King Solamona, the 'Law-giver of our Nation'.]

Yee shall understand (my deare Friends), that amongst the Excellent Acts of that King one above all hath the preheminence. It was the Erection and Institution of an Order or Society which wee call

Salomons House, the Noblest Foundation (as wee thinke) that ever was upon the Earth, And the Lanthorne* of this Kingdome. It is dedicated to the Study of the Works and Creatures of God. Some thinke that it beareth the Founders Name a little corrupted, as if it should be Solamona's House. But the Records write it as it is spoken. So as I take it to bee denominate of the King of the Hebrewes, which is famous with you and no Stranger to us. For wee have some Parts of his works which with you are lost, Namely that Naturall History which hee wrote of all Plants, from the Cedar of Libanus to the Mosse that groweth out of the Wall,[8] And of all things that have Life and Motion. This maketh me thinke that our King finding himselfe to Symbolize* in many things with that King of the Hebrewes (which lived many yeares before him), honoured him with the Title of this Foundation. And I am the rather induced to be of this Opinion for that I finde in ancient Records this Order of Societie is sometimes called Salomons House, And sometimes the Colledge of the sixe Daies Workes: wherby I am satisfied That our Excellent King had learned from the Hebrewes That God had created the World and all that therin is within sixe Dayes, And therefore hee instituting that House for the finding out of the true Nature of all Things (wherby God mought have the more Glory in the Workemanship of them, and Men the more fruit in the use of them), did give it also that second Name.

But now to come to our present purpose. When the King had forbidden to all his People Navigation into any Part that was not under his Crowne, he made neverthelesse this Ordinance: That every twelve yeares ther should be set forth out of this Kingdome two Ships, appointed to severall Voyages; That in either of these Shipps ther should be a Mission of three of the Fellowes or Brethren of Salomons House, whose Errand was onely to give us Knowledge of the Affaires and State of those Countries to which they were designed, And especially of the Sciences, Arts, Manufactures, and Inventions of all the World; And withall to bring unto us Bookes, Instruments, and Patternes in every kind: .... Thus you see wee maintaine a Trade not for Gold, Silver, or Iewels, Nor for Silkes, Nor for Spices, Nor any other Commodity of Matter, But onely for Gods first Creature, which was Light: To have Light (I say) of the Growth of all Parts of the World ....

[As a great honour, one of the Fathers of Salomon's House comes to tell the travellers about this institution and its scientific researches.]

Francis Bacon (1561–1626)

He was a Man of middle Stature and Age, comely of Person, and had an Aspect as if he pittied Men ... We found him in a faire Chamber, richly hanged, and carpetted under Foote, without any Degrees* to the State*. He was sett upon a Low Throne richly adorned, and a rich Cloath of State over his Head of blew Sattin Embroidered. He was alone, save that he had two Pages of Honour, on either Hand one, finely attired in White. His Under Garments were the like that we saw him weare in the Chariott; but in stead of his Gowne he had on him a Mantle with a Cape of the same fine Black fastned about him. When we came in, as we were taught, we bowed Lowe at our first Entrance, And when we were come neare his Chaire he stood up, holding forth his Hand ungloved, and in Posture of Blessing, And we every one of us stooped downe and kissed the Hemme of his Tippett.* That done, the rest departed and I remayned. Then hee warned* the Pages forth of the Roome, and caused mee to sit downe beside him, and spake to me thus in the Spanish Tongue.

God blesse thee, my Sonne; I will give thee the greatest Iewell I have: For I will impart unto thee, for the Love of God and Men, a Relation of the true State of Salomons House. Sonne, to make you know the true state of Salomons House I will keepe this order. First I will set forth unto you the End of our Foundation. Secondly the Preparations and Instruments we have for our Workes. Thirdly the severall Employments and Functions wherto our Fellowes are assigned. And fourthly the Ordinances and Rites which we observe.

The End of our Foundation is the Knowledge of Causes and Secrett Motions of Things, And the Enlarging of the bounds of Humane Empire to the Effecting of all Things possible.

The Preparations and Instruments are these. We have large and deepe Caves of severall Depths. The deepest are sunke 600 Fathome, And some of them are digged and made under great Hills and Mountaines, So that if you reckon together the Depth of the Hill and the Depth of the Cave they are (some of them) above three Miles deepe. For wee finde that the Depth of a Hill and the Depth of a Cave from the Flat is the same Thing, Both remote alike from the Sunn and Heavens Beames, and from the Open Aire. These Caves we call the Lower Region; And wee use them for all Coagulations,* Indurations,* Refrigerations, and Conservations of Bodies. We use them likewise for the Imitation of Naturall Mines, And the Producing also of New Artificiall Mettalls by Compositions and Materialls which we use, and lay ther for many yeares. Wee use them also sometimes (which may seeme strange) for Curing of some Diseases, and for Prolongation of Life in some Hermits that choose to live ther, well accommodated of all things necessarie, and indeed live very long; By whom also we learne many things.

We have Burialls in severall Earths, wher we put diverse Cements, as the Chineses doe their Porcellane. But we have them in greater Varietie and some of them more fine. We have also great variety of Composts and Soiles for the Making of the Earth Fruitfull.

We have High Towers, the Highest about halfe a Mile in Height, And some of them likewise set upon High Mountaines, So that the Vantage of the Hill with the Tower is in the highest of them three Miles at least. And these Places wee call the Upper Region, Accounting the Aire betweene the High Places and the Lowe as a Middle Region. Wee use these Towers, according to their severall Heights and Situations, for Insolation, Refrigeration, Conservation; And for the View of divers Meteors, As Windes, Raine, Snow, Haile, And some of the Fiery Meteors also. And upon them in some Places are Dwellings of Hermits, whom wee visit sometimes and instruct what to observe.

We have great Lakes, both Salt and Fresh; wherof we have use for the Fish and Fowle. We use them also for Burialls of some Naturall Bodies: For we finde a Difference in Things buried in Earth or in Aire below the Earth, and things buried in Water. We have also Pooles, of which some doe straine Fresh Water out of Salt, And others by Art doe turne Fresh Water into Salt. We have also some Rocks in the Midst of the sea, And some Bayes upon the Shore for some Works wherin is required the Ayre and Vapour of the Sea. We have likewise Violent Streames and Cataracts, which serve us for many Motions: And likewise Engines for Multiplying and Enforcing of Windes, to set also on going diverse Motions.

We have also a Number of Artificiall Wels and Fountaines, made in Imitation of the Naturall Sources and Baths, As tincted* upon Vitrioll,* Sulphur, Steele, Brasse, Lead, Nitre and other Mineralls. And againe wee have little Wells for Infusions of many Things, wher the Waters take the Vertue* quicker and better than in Vessells or Basins. And amongst them we have a Water which wee call Water of Paradise, being, by that we doe to it, made very Soveraigne for Health and Prolongation of Life.

We have also Great and Spatious Houses, wher we imitate and demonstrate* Meteors, As Snow, Haile, Raine, some Artificiall Raines of Bodies, and not of Water, Thunders, Lightnings; Also Generations of Bodies in Aire, As Froggs, Flies, and diverse Others.

We have also certaine Chambers which wee call Chambers of

Health, wher wee qualifie* the Aire as we thinke good and proper for the Cure of diverse Diseases and Preservation of Health.

Wee have also faire and large Baths of severall Mixtures, for the Cure of Diseases and the Restoring of Mans Body from Arefaction:* And Others for the Confirming of it in Strength of Sinnewes, Vitall Parts, and the very Iuyce and Substance of the Body.

We have also large and various Orchards and Gardens; Wherin we do not so much respect Beauty as Variety of Ground and Soyle, proper for diverse Trees and Herbs: And some very spatious, wher Trees and Berries are set, wherof we make diverse Kinds of Drinks, besides the Vineyards. In these wee practise likewise all Conclusions of Grafting and Inoculating,* as well of Wilde-Trees as Fruit-Trees, which produceth many Effects. And we make (by Art) in the same Orchards and Gardens, Trees and Flowers to come earlier or later than their Seasons; And to come up and beare more speedily than by their Naturall Course they doe. We make them also by Art greater much than their Nature; And their Fruit greater and sweeter, and of differing Tast, Smell, Colour, and Figure from their Nature. And many of them we so Order as they become of Medicinall Use.

Wee have also Meanes to make diverse Plants rise by Mixtures of Earths without Seedes; And likewise to make diverse New Plants differing from the Vulgar, and to make one Tree or Plant turne into another.

We have also Parks and Enclosures of all Sorts, of Beasts and Birds; which wee use not onely for View or Rarenesse but likewise for Dissections, and Triells, That thereby we may take light what may be wrought upon the Body of Man. Wherin we finde many strange Effects, As Continuing Life in them though diverse Parts which you account Vitall be perished and taken forth, Resussitating of some that seeme Dead in Appearance, And the like. We try also all Poysons and other Medicines upon them, as well of Chyrurgery as Phisicke.* By Art likewise we make them Greater or Taller than their Kinde is, And contrary-wise Dwarfe them and stay their Grouth. Wee make them more Fruitfull and Bearing than their Kind is, and contrary-wise Barren and not Generative. Also we make them differ in Colour, Shape, Activity many wayes. We finde Meanes to make Commixtures* and Copulations of diverse Kindes, which have produced many New Kindes, and them not Barren as the generall Opinion is. We make a Number of Kindes of Serpents, Wormes, Flies, Fishes, or Putrefaction, Wherof some are advanced (in effect) to be Perfect Creatures, like Beastes or Birds, And have Sexes and doe Propagate. Neither doe we this by Chance, but wee know before hand of what Matter and Commixture what Kinde of those Creatures will arise.

Wee have also Particular Pooles, wher we make Trialls upon Fishes as we have said before of Beasts and Birds.

Wee have also Places for Breed and Generation of those Kindes of Wormes and Flies which are of Speciall Use; Such as are with you your Silkwormes and Bees.

I will not hold you long with recounting of our Brew-Howses, Bake-Howses, and Kitchins, wher are made diverse Drinks, Breads, and Meats, Rare and of speciall Effects. Wines we have of Grapes, and Drinkes of other Iuyce, of Fruits, of Graines and of Rootes, And of Mixtures with Honey, Sugar, Manna, and Fruits dryed and decocted; Also of the Teares or Wounding of Trees, And of the Pulp of Canes. And these Drinkes are of severall Ages, some to the Age or Last of fourtie yeares. We have Drinks also brewed with severall Herbs, and Roots, and Spices; Yea with severall Fleshes and White-Meates, Wherof some of the Drinkes are such as they are in effect Meat and Drinke both: So that Diverse, especially in Age, doe desire to live with them, with little or no Meate or Bread. And above all wee strive to have Drinks of Extreame Thin Parts, to insinuate into the Body and yet without all Biting, Sharpenesse, or Fretting,* Insomuch as some of them, put upon the Back of your Hand will, with a little stay, passe through to the Palme and taste yet Milde to the Mouth. Wee have also Waters which we ripen in that fashion as they become Nourishing, So that they are indeed excellent Drinke, And Many will use no other. Breads we have of severall Graines, Roots, and Kernells; Yea and some of Flesh and Fish, Dryed, With diverse kindes of Leavenings* and Seasonings, So that some doe extreamely move Appetites; Some doe Nourish so as diverse doe live of them without any other Meate, Who live very long. So for Meates wee have some of them so beaten and made tender and mortified,* yet without all Corrupting, as a Weake Heate of the Stomach will turne them into good Chylus,* As well as a Strong Heate would Meate otherwise prepared. We have some Meates also, and Breads and Drinks, which taken by Men enable them to Fast long after; And some other that used make the very Flesh of Mens Bodies sensibly more Hard and Tough, And their Strength farre greater than otherwise it would bee.

Wee have Dispensatories, or Shops of Medicines. Wherin you may easely thinke, if we have such Variety of Plants and Living Creatures, more than you have in Europe (for we know what you have), the Simples*, Druggs, and Ingredients of Medicines must likewise be in so much the greater Variety. Wee have them likewise of diverse Ages, and long Fermentations*. And for their Preparations, wee have not onely all Manner of Exquisite Distillations and Separations,* and especially by Gentle Heates and Percolations through diverse Strainers, yea and Substances, But also exact Formes of Composition* wherby they incorporate* allmost as they were Naturall Simples.

Wee have also diverse Mechanicall Arts which you have not, And Stuffes made by them, As Papers, Linnen, Silks, Tissues; dainty Works of Feathers of wonderfull Lustre; excellent Dies, and many others. And Shops likewise, as well for such as are not brought into Vulgar use amongst us as for those that are. For you must know that of the Things before recited many of them are growne into use throughout the Kingdome; But yet, if they did flow from our Invention wee have of them also for Patternes and Principalls.

Wee have also Fournaces of great Diversities, and that keepe great Diversitie of Heates: Fierce and Quicke; Strong and Constant; Soft and Milde; Blowne,* Quiet, Dry, Moist, And the like. But above all we have Heates, in Imitation of the Sunnes and Heavenly Bodies Heates, that passe diverse Inequalities, and (as it were) Orbs,* Progresses and Returnes, wherby we produce admirable effects. Besides wee have Heates of Dungs; and of Bellies and Mawes* of Living Creatures, and of their Blouds and Bodies; and of Hayes and Herbs layd up moist; of Lime unquenched; and such like. Instruments also which generate Heate onely by Motion. And further, Places for Strong Insolations; And againe Places under the Earth, which by Nature or Art yeeld Heate. These diverse Heates wee use As the Nature of the Operation which wee intend requireth.

Wee have also Perspective-Houses, wher wee make Demonstrations of all Lights and Radiations, And of all Colours; And out of Things uncoloured and Transparent wee can represent unto you all severall Colours, Not in Raine-Bowes (as it is in Gemms and Prismes) but of themselves Single. Wee represent also all Multiplications of Light, which wee carry to great Distance and make so Sharp as to discerne small Points and Lines. Also all Colourations of Light, All Delusions and Deceits of the Sight in Figures, Magnitudes, Motions, Colours, All Demonstrations of Shadowes. Wee finde also diverse Meanes yet unknowne to you of Producing of Light originally, from diverse Bodies. Wee procure meanes of Seeing obiects a-farr off, As in the Heaven and Remote Places; And represent Things Neare as A-farr off, And Things A-farr off as Neare, Making Faigned Distances. Wee have also Helps for the Sight, farre above Spectacles and Glasses in use. Wee have also Glasses and Meanes to see Small and

Minute Bodies perfectly and distinctly, As the Shapes and Colours of Small Flies and Wormes, Graines and Flawes in Gemmes which cannot otherwise be seen, Observations in Urine & Bloud not otherwise to be seen. Wee make Artificiall Raine-Bowes, Halo's, and Circles about Light. Wee represent also all manner of Reflexions, Refractions, and Multiplications of Visuall Beames of Objects.

Wee have also Pretious Stones of all kindes, many of them of great Beauty and to you unknowne: Crystalls likewise, And Glasses of diverse kinds, And amongst them some of Mettals Vitrificated,* and other Materialls besides those of which you make Glasse. Also a Number of Fossiles and Imperfect Mineralls which you have not. Likewise Loadstones of Prodigious Vertue,* And other rare Stones, both Naturall and Artificiall.

Wee have also Sound-Houses, wher wee practise and demonstrate all Sounds and their Generation. Wee have Harmonies which you have not, of Quarter-Sounds and lesser Slides* of Sounds. Diverse Instruments of Musick likewise to you unknowne, some sweeter than any you have, Together with Bells and Rings that are dainty and sweet. Wee represent Small Sounds as Great and Deepe, Likewise Great Sounds Extenuate* and Sharpe; Wee make diverse Tremblings and Warblings of Sounds, which in their Originall are Entire. Wee represent and imitate all Articulate Sounds and Letters, and the Voices and Notes of Beasts and Birds. Wee have certaine Helps which, sett to the Eare, doe further the Hearing greatly. Wee have also diverse Strange and Artificiall Eccho's, Reflecting the Voice many times and as it were Tossing it, And some that give back the Voice Lowder than it came, some Shriller and some Deeper; Yea some rendring the Voice Differing in the Letters or Articulate Sound from that they receyve. Wee have also meanes to convey Sounds in Trunks* and Pipes, in strange Lines, and Distances.

Wee have also Perfume-Houses; wherewith we ioyne also Practises of Tast. Wee Multiply* Smells, which may seeme strange. Wee Imitate Smells, making all Smells to breath out of other Mixtures than those that give them. Wee make diverse Imitations of Tast likewise, so that they will deceyve any Mans Tast. And in this House wee containe also a Confiture-House,* wher wee make all Sweet-Meats, Dry and Moist, And diverse pleasant Wines, Milks, Broaths,* and Sallets* farr in greater variety than you have.

Wee have also Engine-Houses,* wher are prepared Engines* and Instruments for all Sorts of Motions. Ther we imitate and practise to make Swifter Motions than any you have, either out of your Musketts or any Engine that you have; And to Make them and Multiply them more Easily, and with Small Force, by Wheeles and other Meanes; And to make them Stronger and more Violent than yours are, Exceeding your greatest Cannons and Basilisks.* Wee represent also Ordnance and Instruments of Warr, and Engines of all Kindes, And likewise New Mixtures and Compositions of Gun-Powder, Wilde-Fires* burning in Water and Unquenchable. Also Fire-workes of all Variety, both for Pleasure and Use. Wee imitate also Flights of Birds; Wee have some Degrees of Flying in the Ayre. Wee have Shipps and Boates for Going under Water, and Brooking* of Seas; Also Swimming-Girdles* and Supporters. Wee have divers curious Clocks And other like Motions of Returne*, And some Perpetuall Motions. Wee imitate also Motions of Living Creatures, by Images of Men, Beasts, Birds, Fishes and Serpents. Wee have also a great Number of other Various Motions, strange for Equality, Finenesse, and Subtilty.

Wee have also a Mathematical House, wher are represented all Instruments, as well of Geometry as Astronomy, exquisitely made.

Wee have also Houses of Deceits of the Senses; wher we represent all manner of Feates of Iugling, False Apparitions, Impostures and Illusions, And their Fallaces*. And surely you will easily beleeve that wee, that have so many Things truely Naturall which induce Admiration, could in a World of Particulars deceive the Senses, if wee would disguise those Things and labour to make them seeme more Miraculous. But wee doe hate all Impostures and Lies: Insomuch as wee have severely forbidden it to all our Fellowes, under paine of Ignominy and Fines, that they doe not shew any Naturall worke or Thing Adorned or Swelling,* but onely Pure as it is, and without all Affectation of Strangenesse.

These are (my Sonne) the Riches of Salomons House.

For the severall Employments and Offices of our Fellowes Wee have Twelve that Sayle into Forraine Countries, under the Names of other Nations (for our owne wee conceale), Who bring us the Books, and Abstracts, and Patternes of Experiments of all other Parts. These wee call Merchants of Light.

Wee have Three that Collect the Experiments which are in all Bookes. These wee call Depredatours.*

Wee have Three that Collect the Experiments of all Mechanicall Arts, And also of Liberall Sciences, And also of Practises which are not Brought into Arts. These we call Mystery-Men.*

Wee have Three that try New Experiments such as themselves thinke good. These wee call Pioners or Miners.

Wee have Three that Drawe the Experiments of the Former Foure into Titles and Tables, to give the better light for the drawing of Observations and Axiomes* out of them. These wee call Compilers.[9]

We have Three that bend themselves, Looking into the Experiments of their Fellowes, and cast about how to draw out of them Things of Use and Practise for Mans life and Knowledge, as well for Workes as for Plaine Demonstration of Causes, Meanes of Naturall Divinations*, and the easie and cleare Discovery of the Vertues and Parts of Bodies. These wee call Dowry-men* or Benefactours.[10]

Then after diverse Meetings and Consults of our whole Number to consider of the former Labours and Collections, wee have Three that take care out of them to Direct New Experiments of a Higher Light, more Penetrating into Nature than the Former. These wee call Lamps.

Wee have Three others that doe Execute the Experiments so Directed, and Report them. These wee call Inoculatours.*

Lastly, wee have Three that raise the former Discoveries by Experiments into Greater Observations, Axiomes, and Aphorismes.* These wee call Interpreters of Nature.

Wee have also, as you must thinke, Novices and Apprentices, that the Succession of the former Employed Men doe not faile, Besides a great Number of Servants and Attendants, Men and Women. And this we doe also: We have Consultations which of the Inventions and Experiences which wee have discovered shall be Published, and which not, And take all an Oath of Secrecy for the Concealing of those which wee thinke fitt to keepe Secrett; Though some of those we doe reveale sometimes to the State, and some not.

For our Ordinances and Rites: Wee have two very Long and Faire Galleries. In one of these wee place Patternes and Samples of all manner of the more Rare and Excellent Inventions; In the other wee place the Statua's of all Principall Inventours. There we have the Statua of your Columbus, that discovered the West-Indies; Also the Inventour of Shipps; Your Monke[11] that was the Inventour of Ordnance and of Gunpowder; The Inventour of Musicke; The Inventour of Letters; the Inventour of Printing; The Inventour of Observations of Astronomy; The Inventour of Works in Mettall; The Inventour of Glasse; The Inventour of Silke of the Worme; The Inventour of Wine; The Inventour of Corne and Bread; The Inventour of Sugars; And all these by more certaine Tradition than you have. Then have we diverse Inventours of our Owne of Excellent Workes, Which since you have not seene it were too long to make Descriptions of them; And besides, in the right Understanding of those Descriptions you might easily erre. For upon every Invention of Valew wee erect a Statua to the Inventour, and give him a Liberall and Honourable Reward. These Statua's are some of Brasse, some of Marble and Touchstone,* some of Cedar and other speciall Woods guilt* and adorned, some of Iron, some of Silver, some of Gold.

We have certaine Hymnes and Services which wee say dayly, of Laud and Thanks to God for his Marveillous Works, And Formes of Prayers imploring his Aide and Blessing for the Illumination of our Labours, and the Turning of them into Good and Holy Uses.

Lastly, wee have Circuites* or Visits of Divers Principall Citties of the Kingdome; wher, as it commeth to passe, we doe publish such New Profitable Inventions as wee thinke good. And wee doe also declare Naturall Divinations* of Diseases, Plagues, Swarmes of Hurtfull Creatures, Scarcety, Tempests, Earthquakes, Great Inundations, Cometts, Temperature of the Yeare, and diverse other Things, And wee give Counsell thereupon what the People shall doe for the Prevention and Remedy of them.

And when Hee had sayd this Hee stood up: And I, as I had beene taught, kneeled downe, and He layd his Right Hand upon my Head and said: God blesse thee, my Sonne, and God blesse this Relation which I have made. I give thee leave to Publish it for the Good of other Nations, For wee here are in Gods Bosome, a Land unknowne. And so hee left mee Having assigned a Valew of about two Thousand Duckets* for a Bounty to mee and my Fellowes. For they give great Largesses* where they come, upon all occasions.

The rest was not Perfected.

Robert Boyle

(1627–1691)

3 Experiments with the air-pump (1660)

Robert Boyle was the youngest son of Richard Boyle, first earl of Cork. He was educated privately, then at Eton, and finally in Geneva, travelling in Italy and elsewhere (he read Galileo's *Dialogue on the Two Chief World Systems* at Florence in 1642). He had his own laboratory for much of his life, and learned chemistry (as he recorded in *The Sceptical Chymist*) practically, from "illiterate persons" ignorant of theory. He was associated with the group of scientists around Samuel Hartlib in the 1640s, and moved to Oxford in 1656, where he came into contact with the Wilkins circle, and employed Robert Hooke as his assistant. He moved to London in 1668, where his house became a private research centre. He was a proponent of Baconian experimentation, but also developed a corpuscular theory of matter. He made important contributions to chemistry, especially on the processes of colour change and its application to chemical classification (such as acid and alkali tests), and wrote voluminously on pneumatics, low-temperature thermometry, colour, and theology.

Once past the preface, in which he apologizes at length for the imperfections of the work and the prolixity of his style, Boyle's first major scientific publication is a faithful embodiment of the proper form for writing up experiments, as laid down by Bacon in the *Parasceve*: no rhetorical flourishes, no superfluous citation of authorities, no verbal ornament, but a straightforward account of the operations performed. Boyle is one of the first exponents of this new literary genre, the written-up experiment, of which the conventional parts are: description of the apparatus; account of the phenomena to be investigated; narrative of the experiments; record of the results; theoretical implications for further experiments. Like so many of the Bacon-influenced scientists in the Royal Society Boyle distrusted "theory", and remained excessively cautious about formulating hypotheses of scientific laws. Yet his grasp of the fundamental problems of the "Spring" or elasticity of the air, its "weight" (air-pressure), and the effects of a vacuum, are visible throughout the experimental record. Although well received by many scientists, English and European, Boyle was attacked by two traditional critics, Franciscus Linus and Thomas Hobbes. These he answered in an impressively argued *Defence of the Doctrine touching the Spring and Weight of the Air* (1662), which contained the first formulation of Boyle's Law, namely that at constant temperature gas volume and pressure are inversely propor-

45

tional. This law was confirmed in two *Continuations* of his experiments, in 1669 (*Works*, ed. Birch, vol. III, pp. 175–287) and 1692 (IV, pp. 505–93).

Source: *New Experiments Physico-Mechanical, Touching the Spring of the Air, and its Effects; Made, for the most Part, in a New Pneumatical Engine. Written by Way of Letter to the Right Honourable Charles Lord Viscount of Dungarvan* ... (1660), in Boyle's *Works*, ed. Thomas Birch, 6 vols. (second edition; London, 1772) I, pp. 1–117; the *Defence* follows (I, pp. 118–242). The "New Pneumatical Engine" is the air-pump devised for Boyle by Robert Hooke. The work is cast in the form of a letter to Boyle's nephew, a sign that he had not yet moved from the "literary" to the "scientific" mode.

Preface

Of my being somewhat prolix in many of my experiments I have these reasons to render: that some of them being altogether new seemed to need the being circumstantially related to keep the reader from distrusting them; that divers circumstances I did here and there set down for fear of forgetting them when I may hereafter have occasion to make use of them in my other writings; that in divers cases I thought it necessary to deliver things circumstantially, that the person I addressed them to might, without mistake, and with as little trouble as is possible be able to repeat such unusual experiments; and that after I consented to let my observations be made public, the most ordinary reason of my prolixity was that, foreseeing that such a trouble as I met with in making those trials carefully, and the great expense of time that they necessarily require (not to mention the charges of making the engine and employing a man to manage it) will probably keep most men from trying again these experiments, I thought I might do the generality of my readers no unacceptable piece of service by so punctually* relating what I carefully observed, that they may look upon these narratives as standing records in our new pneumatics, and need not reiterate themselves an experiment to have as distinct an idea of it as may suffice them to ground their reflexions and speculations upon.

And because sometimes 'tis the discourse made upon the experiment that makes it appear prolix, I have commonly left a conspicuous interval betwixt such discourses and the experiments whereunto they belong or are annexed; that they who desire only the historical* part of the account we give of our engine* may read the narratives without being put to the trouble of reading the reflexions too; which I here take notice of for the sake of those that are well versed in the new philosophy and in the mathematicks, that such may skip what was designed but for such persons as may be less acquainted, even than I, with matters of this nature (scarce so much as mentioned by any writer in our language) and not for them from whom I shall be much more forward to learn than to pretend to teach them. Of my being wont to speak rather doubtfully or hesitantly than resolvedly, concerning matters wherein I apprehend some difficulty, I have in another treatise[1] (which may, through God's assistance, come abroad ere long) given a particular and, I hope, a satisfactory account; wherefore I shall now defend my practice but by the observation of *Aristotle*, who somewhere notes that to seem to know all things certainly and to speak positively of them is a trick of bold and young fellows, whereas those that are indeed intelligent and considerate are wont to employ more wary or diffident expressions.

There are divers reflexions, and other passages in the following epistle and even some experiments (occasionally mentioned) which may seem either impertinent* or superfluous, but are not so; being purposely written either to evince some truth opposed, or disprove some erroneous conceit maintained by some eminent new philosopher or by some other ingenious men, who, I presumed, would easily forgive me the having on such occasions purposely omitted their names; though an inquisitive person will probably discover divers of them by the mention of the opinions disprov'd in the experiments I am excusing.

Ever since I discerned the usefulness of speculative geometry to natural philosophy the unhappy distempers of my eyes have so far kept me from being much conversant in it that I fear I shall need the pardon of my mathematical readers for some passages which, if I had been deeply skill'd in geometry, I should have treated more accurately ....

But if it be demanded, why then I did not make it fitter for the press before I sent it thither? my answer must be that not at first imagining that this sort of experiments would prove any thing near so troublesome either to make or to record as I afterwards found them, I did, to engage the printer to dispatch,* promise him to send him the whole epistle in a very short time. So that although now and then the occasional vacations of the press, by reason of festivals or the absence of the corrector, gave me the leisure to expatiate upon some subject; yet being oftentimes called upon to dispatch the papers to the press, my promise, and many unexpected avocations, obliged me to a haste which, though it hath detracted nothing from the faithfulness of the historical* part of our book, hath (I fear) been disadvantageous enough to all the rest ....

Robert Boyle (1627–1691)

New Experiments Physico-Mechanical

... And I am not faintly induced to make choice of this subject rather than any of the expected chymical ones to entertain your Lordship upon, by these two considerations: the one, that the air being so necessary to human life that not only the generality of men but most other creatures that breathe cannot live many minutes without it, any considerable discovery of its nature seems likely to prove of moment to mankind. And the other is that the ambient air being that whereto both our own bodies and most of the others we deal with here below are almost perpetually contiguous, not only its alterations have a notable and manifest share in those obvious effects that men have already been invited to ascribe thereunto (such as are the various distempers incident to human bodies, especially if crazy in the spring, the autumn, and also on most of the great and sudden changes of weather), but likewise the further discovery of the nature of the air will probably discover to us that it concurs more or less to the exhibiting of many phænomena in which it hath hitherto scarce been suspected to have any interest.* So that a true account of any experiment that is new concerning a thing wherewith we have such constant and necessary intercourse may not only prove of some advantage to human life, but gratify philosophers by promoting their speculations on a subject which hath so much opportunity to solicit their curiosity.

> [Boyle records how he was inspired to construct an air-pump by the publication (in Caspar Schott's *Mechanica hydraulico-pneumatica*, 1657) of the invention of Otto Guericke (1602–1686), Mayor of Magdeburg. Finding various inconveniences in this first model, Boyle asked Robert Hooke to make a better one.]

To give your Lordship then, in the first place, some account of the engine itself: it consists of two principal parts, a glass vessel, and a pump to draw the air out of it.

The former of these (which we, with the glass-men, shall often call a receiver, for its affinity to the large vessels of that name used by chymists) consists of a glass with a wide hole at the top, of a cover to that hole, and of a stop-cock fastened to the end of the neck at the bottom.

The shape of the glass you will find expressed in the first figure of the annexed scheme.* And for the size of it, it contained about 30 wine quarts, each of them containing near two pound (of 16 ounces to the pound) of water. We should have been better pleased with a more capacious vessel, but the glass-men professed themselves unable to blow a larger of such a thickness and shape as was requisite to our purpose.

At the very top of the vessel A you may observe a round hole, whose diameter B C is of about four inches; and whereof the orifice is incircled with a lip of glass almost an inch high: for the making of which lip it was requisite (to mention that upon the by, in case your Lordship should have such another engine made for you) to have a hollow and tapering pipe of glass drawn out, whereof the orifice above mentioned was the basis; and then to have the cone cut off with an hot iron within about an inch of the points B, C.

The use of the lip is to sustain the cover delineated in the second figure; where D E points out a brass ring, so cast as that it doth cover the lip B C of the first figure and is cemented on, upon it, with a strong and close cement. To the inward tapering orifice of this ring (which is about three inches over) are exquisitely* ground the sides of the brass stopple F G, so that the concave superficies of the one, and the convex of the other, may touch one another in so many places as may leave as little access as possible to the external air. And in the midst of this cover is left a hole H I, of about half an inch over, invironed also with a ring or socket of the same metal and fitted likewise with a brass stopple K, made in the form of the key of a stop-cock, and exactly ground into the hole H I it is to fill; so as that, though it be turned round in the cavity it possesses it will not let in the air, and yet may be put in or taken out at pleasure for uses to be hereafter mentioned. In order to some of which it is perforated with a little hole 8, traversing the whole thickness of it at the lower end; through which, and a little brass ring L fastened to one side (no matter which) of the bottom of the stopple F G, a string 8, 9, 10, might pass, to be employed to move some things in the capacity of the emptied vessel without any where unstopping it.

The last thing belonging to our receiver is the stop-cock, designed in the first figure by N, for the better fastening of which to the neck and exacter exclusion of the air there was sodered* on to the shank of the cock X a plate of tin M T U W, long enough to cover the neck of the receiver. But because the cementing of this was a matter of some difficulty it will not be amiss to mention here the manner of it; which was, that the cavity of the tin plate was filled with a melted cement, made of pitch, rosin* and wood-ashes, well incorporated*; and to hinder this liquid mixture from getting into the orifice Z of the shank X that hole was stopped with a cock, to which was fastened a string,

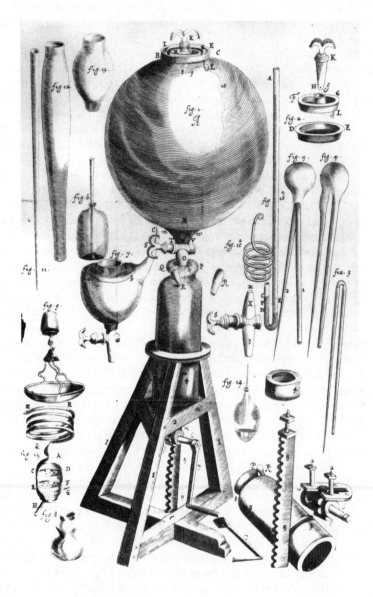

Plate 1   Boyle's air-pump and related equipment; from Boyle, *Nova Experimenta Physico-mechanica vi aeris elastica* (Geneva, 1661)

whereby it might be pulled out at the upper orifice of the receiver; and then the glass neck of the receiver being well warmed was thrust into this cement, and over the shank, whereby it was effected that all the space betwixt the tin plate and the receiver, and betwixt the internal superficies of the receiver and the shank of the cock was filled with the cement. And so we have dispatched the first upper part of the engine.

The undermost remaining part consists of a frame and of a sucking-pump or, as we formerly called it, an air-pump, supported by it . . . .

[Boyle gives further particular instructions as to the correct operating of the air-pump.]

EXPERIMENT I

To proceed now to the phænomena exhibited to us by the engine above described; I hold it not unfit to begin with what doth constantly and regularly offer itself to our observation, as depending upon the fabric* of the engine itself and not upon the nature of this or that particular experiment, which it is employed to try.

First then, upon the drawing down of the sucker (the valve being shut) the cylindrical space deserted* by the sucker is left devoid of air; and therefore, upon the turning of the key the air contained in the receiver rusheth into the emptied cylinder, till the air in both these vessels be brought to about an equal measure of dilatation. And therefore, upon shutting the receiver by returning the key, if you open the valve and force up the sucker again you will find that after this first exsuction* you will drive out almost a whole cylinder full of air; but at the following exsuctions you will draw less and less of air out of the receiver into the cylinder, because there will still remain less and less air in the receiver itself, and consequently the particles of the remaining air, having more room to extend themselves in, will less press out one another. This you will easily perceive by finding that you still force less and less air out of the cylinder, so that when the receiver is almost exhausted you may force up the sucker almost to the top of the cylinder before you will need to unstop the valve to let out any air. And if at such time, the valve being shut, you let go the handle of the pump, you will find the sucker forcibly carried up to the top of the cylinder by the protrusion of the external air; which, being much less rarified than that within the cylinder, must have a more forcible pressure upon the sucker than the internal is able to resist, and by this means you may know how far you have emptied the receiver. And to this we may add on this occasion, that constantly upon the turning of the key to let out the air from the receiver into the emptied cylinder there is immediately produced a considerably brisk noise, especially whilst there is any plenty of air in the receiver.

For the more easy understanding of the experiments triable by our engine I thought it not superfluous nor unseasonable in the recital of this first of them to insinuate that notion by which it seems likely that most, if not all of them, will prove explicable. Your Lordship will easily suppose that the notion I speak of is that there is a spring or elastical power in the air we live in. By which ἐλατὴρ[2] or spring of the air that which I mean is this, that our air either consists of, or at least abounds with parts of such a nature that in case they be bent or compressed by the weight of the incumbent part of the atmosphere or by any other body, they do endeavour, as much as in them lieth, to free themselves from that pressure by bearing against the contiguous bodies that keep them bent; and as soon as those bodies are removed, or reduced to give them way, by presently unbending and stretching out themselves either quite, or so far forth as the contiguous bodies that resist them will permit, and thereby expanding the whole parcel of air these elastical bodies compose.

This notion may perhaps be somewhat further explained by conceiving the air near the earth to be such a heap of little bodies, lying one upon another, as may be resembled to a fleece of wool.[3] For this (to omit other likeness betwixt them) consists of many slender and flexible hairs, each of which may indeed, like a little spring, be easily bent or rolled up but will also, like a spring, be still endeavouring to stretch itself out again. For though both these hairs and the aëreal corpuscles to which we liken them do easily yield to external pressures, yet each of them (by virtue of its structure) is endowed with a power or principle of self-dilation, by virtue whereof, though the hairs may by a man's hand be bent and crouded closer together, and into a narrower room than suits best with the nature of the body, yet, whilst the compression lasts there is in the fleece they compose an endeavour outwards, whereby it continually thrusts against the hand that opposes its expansion. And upon the removal of the external pressure, by opening the hand more or less, the compressed wool doth, as it were, spontaneously expand or display* itself towards the recovery of its former more loose and free condition, till the fleece hath either regained its former dimensions, or at least approached them as near as the compressing hand (perchance not quite opened) will permit. This power of self-dilatation is somewhat more conspicuous in a dry spunge compressed than in a fleece of wool. But yet we rather chose to employ the latter on this occasion because it is not, like a spunge, an entire body but a number of slender and flexible bodies loosely complicated, as the air itself seems to be.

There is yet another way to explicate the spring of the air; namely, by supposing with that most ingenious gentleman, Monsieur *Des Cartes*,[4] that the air is nothing but a congeries or heap of small and (for the most part) of flexible particles, of several sizes and of all kind of figures, which are raised by heat (especially that of the sun) into that fluid and subtle ethereal body that surrounds the earth, and by the restless agitation of that celestial matter wherein those particles swim, are so whirled round that each corpuscle endeavours to beat off all others from coming within the little sphere requisite to its motion about its own centre; and in case any, by intruding into that sphere shall oppose its free rotation, to expel or drive it away: so that, according to this doctrine it imports very little whether the particle of the air have the structure requisite to springs, or be of any other form (how irregular soever) since their elastical power is not made to depend upon their shape or structure but upon the vehement agitation and (as it were) brandishing motion which they receive from the fluid æther that swiftly flows between them, and whirling about each of them (independently from the rest) not only keeps those slender aëreal bodies separated and stretched out (at least, as far as the neighbouring ones will permit), which otherwise, by reason of their flexibleness and weight would flag or curl; but also makes them hit against and knock away each other, and consequently require more room than that which, if they were compressed, they would take up.

By these two differing ways, my Lord, may the springs of the air be explicated. But though the former of them be that which by reason of its seeming somewhat more easy I shall for the most part make use of in the following discourse, yet am I not willing to declare peremptorily for either of them against the other. And indeed, though I have in another treatise endeavoured to make it probable that the returning of elastical bodies (if I may so call them) forcibly bent to their former position may be mechanically* explicated, yet I must confess that to determine whether the motion of restitution* in bodies proceed from this, that the parts of a body of a peculiar structure are put into motion by the bending of the spring, or from the endeavour of some subtle ambient* body whose passage may be opposed or obstructed, or else its pressure unequally resisted by reason of the new shape or magnitude which the bending of a spring may give the pores of it: to determine this, I say, seems to me a matter of more difficulty than at first sight one would easily imagine it. Wherefore I shall decline meddling with a subject which is much more hard to be explicated than necessary to be so by him whose business it is not, in this letter, to assign the adequate cause of the spring of the air but only to manifest that the air hath a spring, and to relate some of its effects.

I know not whether I need annex that though either of the above-mentioned hypotheses, and perhaps some others, may afford us an account plausible enough of the air's spring, yet I doubt whether any of them gives us a sufficient account of its nature. And of this doubt I might here mention some reasons but that, peradventure, I may (God permitting) have a fitter occasion to say something of it elsewhere. And therefore I should now proceed to the next experiment but that I think it requisite, first, to suggest to your Lordship what comes into my thoughts by way of answer to a plausible objection which I foresee you may make against our proposed doctrine touching the spring of the air. For it may be alleged that though the air were granted to consist of springy particles (if I may so speak) yet thereby we could only give an account of the dilatation of the air in wind-guns* and other pneumatical engines, wherein the air hath been compressed and its springs violently bent by an apparent external force, upon the removal of which it is no wonder that the air should, by the motion of restitution, expand itself till it hath recovered its more natural dimensions; whereas, in our above-mentioned first experiment and in almost all others triable in our engine, it appears not that any compression of the air preceded its spontaneous dilatation or expansion of itself. To remove this difficulty I must desire your Lordship to take notice that, of whatever nature the air very remote from the earth may be, and whatever the schools* may confidently teach to the contrary, yet we have divers experiments to evince that the atmosphere we live in is not (otherwise than comparatively to more ponderous bodies) light but heavy. And did not their gravity hinder them it appears not why the streams of the terraqueous* globe, of which our air in great part consists, should not rise much higher than the refractions of the sun and other stars give men ground to think that the atmosphere (even in the judgment of those recent astronomers who seem willing to enlarge its bounds as much as they dare) doth reach.

But lest you should expect my seconding this reason by experience, and lest you should object that most of the experiments that have been proposed to prove the gravity of the air have been either barely proposed or perhaps not accurately tried, I am content, before I pass further, to mention here that I found a dry lamb's bladder containing near about two thirds of a pint, and compressed by a packthread tied about it, to lose a grain and the eighth part of a grain of its former weight by the recess of the air upon my having prickt it; and this with a pair of scales which, when the full bladder and the correspondent weight were in it, would manifestly turn either way with the 32^d part of the grain. And if it be further objected that the air in the bladder was violently compressed by the packthread and the sides of the bladder, we might probably (to waive prolix answers) be furnished with a reply by setting down the differing weight of our receiver when emptied, and when full of uncompressed air, if we could here procure scales fit for so nice an experiment, since we are informed that in the German experiment[5] commended at the beginning of this letter the ingenious triers of it found that their glass vessel, of the capacity of 32 measures, was lighter when the air had been drawn out of it than before, by no less than one ounce and $\frac{3}{10}$, that is, an ounce and very near a third. But of the gravity of the air we may elsewhere have occasion to make further mention.[6]

Taking it then for granted that the air is not devoid of weight, it will not be uneasy* to conceive that that part of the atmosphere wherein we live, being the lower part of it, the corpuscles that compose it are very much compressed by the weight of all those of the like nature that are directly over them; that is, of all the particles of air that, being piled up upon them, reach to the top of the atmosphere. And though the height of this atmosphere, according to the famous *Kepler*[7] and some others, scarce exceeds eight common miles, yet other eminent and later astronomers would promote the confines of the atmosphere to exceed six or seven times that number of miles. And the diligent and learned *Ricciolo*[8] makes it probable that the atmosphere may, at least in divers places, be at least fifty miles high. So that, according to a moderate estimate of the thickness of the atmosphere, we may well suppose that a column of air of many miles in height, leaning upon some springy corpuscles of air here below, may have weight enough to bend their little springs and keep them bent: as (to resume our former comparison) if there were fleeces of wool piled up to a mountainous height one upon another, the hairs that compose the lowermost locks which support the rest would, by the weight of all the wool above them, be as well strongly compressed as if a man should squeeze them together in his hands, or employ any such other moderate force to compress them. So that we need not wonder that upon the taking off the incumbent* air from any parcel of the atmosphere here below, the corpuscles whereof that undermost air consists should display themselves, and take up more room than before.

And if it be objected that in water the weight of the upper and of the lower part is the same, I answer that (besides that it may be well doubted whether the observation, by reason of the great difficulty, hath been exactly made) there is a manifest disparity betwixt the air and water. For I have not found, upon an experiment purposely made (and in another treatise recorded), that water will suffer any considerable compression, whereas we may observe in wind-guns (to mention now no other engines) that the air will suffer itself to be crouded into a comparatively very little room; insomuch that a very diligent examiner of the phænomena of wind-guns would have us believe that in one of them, by condensation, he reduced the air into a space at least eight times narrower than it before possessed. And to this, if we add a noble phænomenon of the experiment *de vacuo*, these things put together may for the present suffice to countenance our doctrine. For that noble experimenter Monsieur *Pascal* (the son) had the commendable curiosity to cause the Torricellian experiment to be tried at the foot, about the middle, and at the top of that high mountain (in *Auvergne*, if I mistake not) commonly called *Le Puy de Domme*; whereby it was found that the mercury in the tube fell down lower, about three inches, at the top of the mountain than at the bottom.[9] And a learned man awhile since informed me that a great Virtuoso,* friend to us both, hath, with not unlike success, tried the same experiment in the lower and upper parts of a mountain in the west of *England*. Of which the reason seems manifestly enough to be this, that upon the tops of high mountains the air which bears against the restagnant* quicksilver is less pressed by the less ponderous incumbent air; and consequently is not able totally to hinder the descent of so tall and heavy a cylinder of quicksilver as at the bottom of such mountains did but maintain an aequilibrium with the incumbent atmosphere ....

EXPERIMENT IV.

Having thus taken notice of some of the constant phænomena of our engine itself, let us now proceed to the experiments triable in it.

We took then a lamb's bladder large, well dried and very limber,* and leaving in it about half as much air as it could contain we caused the neck of it to be strongly tied, so that none of the included air, though by pressure, could get out. This bladder being conveyed into the receiver, and the cover luted* on, the pump was set on work, and after two or three exsuctions of the ambient air (whereby the spring of that which remained in the glass was weakened) the imprisoned air began to swell in the bladder; and as more and more of the air in the receiver was, from time to time, drawn out so did that in the bladder more and more expand itself, and display the folds of the formerly flaccid bladder: so that before we had exhausted the receiver near so much as we could the bladder appeared as full and stretched as if it had been blow up with a quill.*

And that it may appear that this plumpness of the bladder proceeded from the surmounting* of the debilitated* spring of the ambient air remaining in the vessel by the stronger spring of the air remaining in the bladder, we returned the key of the stop-cock, and by degrees allowed the external air to return into the receiver: whereupon it happened, as was expected, that as the air came in from without the distended air in the bladder was proportionably compressed into a narrower room and the sides of the bladder grew flaccid, till the receiver having re-admitted its wonted quantity of air the bladder appeared as full of wrinkles and cavities as before . . . .

EXPERIMENT V.

To try then at once both what it was that expanded the bladder, and what a powerful spring there is even in the air we are wont to think uncompressed, we caused a bladder dry, well tied and blown moderately full, to be hung in the receiver by one end of a string whose other end was fastened to the inside of the cover, and upon drawing out the ambient air that pressed on the bladder the internal air, not finding the wonted resistance, first swelled and distended the bladder, and then broke it with so wide and crooked a rent* as if it had been forcibly torn asunder with hands. After which a second bladder being conveyed in the experiment was repeated with like success: and I suppose it will not be imagined that in this case the bladder was broken by its own fibres, rather than by the imprisoned air.

And of this experience these two phænomena may be taken notice of: the one, that the bladder at its breaking gave a great report, almost like a cracker*; and the other, that the air contained in the bladder had the power to break it with the mentioned impetuosity* long before the ambient air was all, or near all, drawn out of the receiver.

But, to verify what we say in another discourse,[10] where we show that even true experiments may, by reason of the easy mistake of some unheeded circumstance, be unsuccessfully tried, we will advertise on this occasion that we did oftentimes in vain try the breaking of bladders after the manner above mentioned; of which the cause appeared to be this, that the bladders we could not break, having been brought us already blown from those that sold them, were grown dry before they came to our hands, whence it came to pass that if we afterwards tied them very hard they were apt to fret,* and so become unserviceable; and if we tied them but moderately hard their stiffness kept them from being closed so exactly, but that when the included air had in the exhausted receiver distended them as much as easily it could it would in part get out between the little wrinkles of the sphincter* of the neck. Whence also it usually happened that upon the letting in the air from without the bladders appeared more flaccid and empty than before they were put in; whereas when the bladders were brought us moist from the butchers we could, without injuring them, tie their necks so close that none of the air, once blown in, could get out of them but by violently breaking them.

It would not be amiss on this occasion to point at something which may deserve a more deliberate speculation than we can now afford it; namely, that the elastical power of the same quantity of air may be as well increased by the agitation of the aëreal particles (whether only moving them more swiftly and scattering them, or also extending or stretching them out I determine not) within an every way inclosing and yet yielding body; as displayed by the withdrawing of the air that pressed it without. For we found that a bladder but moderately filled with air and strongly tied, being a while held near the fire, not only grew exceeding turgid and hard but afterwards, being brought nearer to the fire, suddenly broke with so loud and vehement a noise as stunned those that were by, and made us for a while after almost deaf.

EXPERIMENT X.

We took a tallow-candle of such a size that eight of them make about a pound; and having in a very commodious candlestick let it down into the receiver, and so suspended it that the flame burnt almost in the middle of the vessel, we did in some two minutes exactly close it up: and upon pumping very nimbly we found that within little more than half a minute after the flame went out, though the snuff* had been purposely left of that length we judged the most convenient for the lasting of the flame.

But the second time having put in the same candle into the receiver (after it had by the blasts of a pair of bellows been freed from fumes) the flame lasted about two minutes from the time the pumper began to draw out the air; upon the first exsuction whereof the flame seemed to contract itself in all its dimensions. And these things were farther observable, that after the two or three first exsuctions of the air the flame (except at the very top) appeared exceeding blue, and that the flame still receded more and more from the tallow till at length it appeared to possess only the very top of the wick, and there it went out.

The same candle being lighted again was shut into the receiver, to try how it would last there without drawing forth the air, and we found that it lasted much longer than formerly; and before it went out receded from the tallow towards the top of the wick, but not near so much as in the former experiment.

EXPERIMENT XVII.

Proceed we now to the mention of that experiment whereof the satisfactory trial was the principal fruit I promised myself from our engine, it being then sufficiently known that in the experiment *de vacuo* the quicksilver in the tube is wont to remain elevated above the surface of that whereon it leans about 27 digits. I considered that if the true and only reason why the quicksilver falls no lower be that at that altitude the mercurial cylinder in the tube is in an æquilibrium with the cylinder of air supposed to reach from the adjacent mercury to the top of the atmosphere, then if this experiment could be tried out of the atmosphere the quicksilver in the tube would fall down to a level with that in the vessel, since then there would be no pressure upon the subjacent* to resist the weight of the incumbent mercury. Whence I inferred (as easily I might) that if the experiment could be tried in our engine the quicksilver would subside below 27 digits, in proportion to the exsuction of air that should be made out of the receiver. For as when the air is shut into the receiver it doth (according to what hath above been taught) continue there as strongly compressed as it did whilst all the incumbent cylinder of the atmosphere leaned immediately upon it, because the glass wherein it is penned up hinders it to deliver itself, by an expansion of its parts, from the pressure wherewith it was shut up. So if we could perfectly draw the air out of the receiver it would conduce as well to our purpose as if we were allowed to try the experiment beyond the atmosphere.

Wherefore (after having surmounted some little difficulties which occurred at the beginning) the experiment was made after this manner. We took a slender and very curiously blown cylinder of glass of near three foot in length, and whose bore had in diameter a quarter of an inch, wanting a hair's breadth; this pipe being hermetically sealed at one end was at the other filled with quicksilver, care being taken in the filling that as few bubbles as was possible should be left in the mercury. Then the tube, being stopt with the finger and inverted, was opened, according to the manner of the experiment, into a somewhat long and slender cylindrical box (instead of which we now are wont to use a glass of the same form) half filled with quicksilver; and so the liquid metal being suffered to subside, and a piece of paper being pasted on level with its uppper surface, the box and tube and all were by strings carefully let down into the receiver; and then, by means of the hole formerly mentioned to be left in the cover, the said cover was slipt along as much of the tube as reached above the top of the receiver, and the interval left betwixt the sides of the hole and those of the tube was very exquisitely* filled up with melted (but not over-hot) diachylon,* and the round chink betwixt the cover and the receiver was likewise very carefully closed up. Upon which closure there approached not any change in the height of the mercurial cylinder, no more than if the interposed glass-receiver did not hinder the immediate pressure of the ambient atmosphere upon the inclosed air; which hereby appears to bear upon the mercury rather by virtue of its spring than of its weight, since its weight cannot be supposed to amount to above two or three ounces, which is inconsiderable in comparison to such a cylinder of mercury as it would keep from subsiding.

All things being thus in a readiness the sucker was drawn down, and immediately upon the egress of a cylinder of air out of the receiver the quicksilver in the tube did, according to expectation, subside: and notice being carefully taken (by a mark fastened to the outside) of the place where it stopt, we caused him that managed the pump to pump again, and marked how low the quicksilver fell at the second exsuction. But continuing this work we were quickly hindered from accurately marking the stages made by the mercury in its descent, because it soon sunk below the top of the receiver, so that we could henceforward mark it no other ways than by the eye. And thus, continuing the labour of pumping for about a quarter of an hour, we found ourselves unable to bring the quicksilver in the tube totally to subside; because when the receiver was considerably emptied of its air, and consequently that little that remained grown unable to resist the irruption* of the external, that air would (in spight of whatever we could do) press in at some little avenue or other, and though much could not thereat get in yet a little was sufficient to counterbalance the pressure of so small a cylinder of quicksilver as then remained in the tube.

Now (to satisfy ourselves farther that the falling of the quicksilver in the tube to a determinate height proceedeth from the æquilibrium wherein it is at that height with the external air, the one gravitating, the other pressing with equal force upon the subjacent mercury) we returned the key and let in some new air. Upon which the mercury immediately began to ascend (or rather to be impelled upwards) in the tube, and continued ascending till, having returned the key, it immediately rested at the height which it had then attained; and so, by turning and returning the key we did several times at pleasure impel it upwards, and check its ascent. And lastly, having given a free egress at the stop-cock to as much of the external air as would come in, the quicksilver was impelled up almost to its first height. I say almost, because it stopt near a quarter of an inch beneath the paper-mark formerly mentioned; which we ascribed to this, that there was (as is usual in this experiment) some little particles of air engaged among those of the quicksilver, which particles, upon the descent of the quicksilver, did manifestly rise up in bubbles towards the top of the tube, and by their pressure, as well as by lessening the cylinder by as much room as they formerly took up in it, hindered the quicksilver from regaining its first height.

This experiment was a few days after repeated in the presence of those excellent and deservedly famous Mathematic Professors, Dr *Wallis*, Dr *Ward*, and Mr *Wren*, who were pleased to honour it with their presence,[11] and whom I name both as justly counting it an honour to be known to them and as being glad of such judicious and illustrious witnesses of our experiment; and it was by their guess that the top of the quicksilver in the tube was defined to be brought within an inch of the surface of that in the vessel . . . .

We formerly mentioned that the quicksilver did not, in its descent, fall as much at a time after the two or three first exsuctions of the air as at the beginning. For, having marked its several stages upon the tube, we found that at the first suck it descended an inch and $\frac{3}{8}$, and at the second an inch and $\frac{3}{8}$, and when the vessel was almost emptied it could scarce at one exsuction be drawn down above the breadth of a barley-corn. And indeed we found it very difficult to measure in what proportion these decrements* of the mercurial cylinder did proceed; partly because (as we have already intimated) the quicksilver was soon drawn below the top of the receiver; and partly because, upon its descent at each exsuction, it would immediately reascend a little upwards, either by reason of the leaking of the vessel at some imperceptible hole or other, or by reason of the motion of restitution in the air, which, being somewhat compressed by the fall as well as weight of the quicksilver, would repel it a little upwards, and make it vibrate a little up and down before they could reduce each other to such an æquilibrium as both might rest in.

But though we could not hitherto make observations accurate enough concerning the measures of the quicksilver's descent to reduce them into any hypothesis, yet would we not discourage any from attempting it; since if it could be reduced to a certainty it is probable that the discovery would not be unuseful.

Lastly, we also observed that if (when the mercury in the tube had been drawn down, and by an ingress permitted to the external air impelled up again to its former height) there were some more air thrust up by the help of the pump into the receiver, the quicksilver in the tube would ascend much above the wonted height of 27 digits, and immediately upon the letting out of that air would fall again to the height it rested at before.

Your Lordship will here perhaps expect that as those who have treated of the Torricellian experiment have for the most part maintained the affirmative or the negative of that famous question, whether or no that noble experiment infer a vacuum? so I should on this occasion interpose my opinion touching that controversy; or at least declare whether or no, in our engine, the exsuction of that air do prove the place deserted by the air sucked out to be truly empty, that is, devoid of all corporeal substance. But besides that I have neither the leisure nor the ability to enter into a solemn debate of so nice* a question; your Lordship may, if you think it worth the trouble, in the Dialogues not long since referred to, find the difficulties on both sides represented, which then made me yield but a very wavering assent to either of the parties contending about the question; nor dare I yet take upon me to determine so difficult a controversy.

EXPERIMENT XXVII.

That the air is the medium whereby sounds are conveyed to the ear hath been for many ages, and is yet the common doctrine of the schools. But this received opinion hath been of late opposed by some philosophers upon the account of an experiment made by the industrious *Kircher*[12] and other learned men; who have (as they assure us) observed that if a bell with a steel clapper be so fastened to the inside of a tube that upon the making the experiment *de vacuo* with that tube the bell remained suspended in the deserted space at the upper end of the tube; and if also a vigorous load-stone* be applied on the outside of the tube to the bell it will attract the clapper, which, upon the removal of the load-stone falling back, will strike against the opposite side of the bell and thereby produce a very audible sound; whence divers have concluded that it is not the air but some more subtle body that is the medium of sounds. But because we conceived that, to invalidate such a consequence from this ingenious experiment (though the most luciferous* that could well be made without some such engine as ours) some things might be speciously* enough alledged, we thought fit to make a trial or two in order to the discovery of what the air doth in conveying of sounds, reserving divers other experiments triable in our engine concerning sounds till we can obtain more leisure to prosecute them. Conceiving it then the best way to make our trial with such a noise as might not be loud enough to make it difficult to discern slighter variations in it, but rather might be both lasting (that we might take notice by what degrees it decreased) and so small that it could not grow much weaker without becoming imperceptible, we took a watch, whose case we opened, that the contained air might have free egress into that of the receiver. And this watch was suspended in the cavity of the vessel only by a pack-thread, as the unlikeliest thing to convey a sound to the top of the receiver; and then closing up the vessel with melted plaister, we listened near the sides of it, and plainly enough heard the noise made by the balance. Those also of us that watched for that circumstance observed that the noise seemed to come directly in a streight line from the watch unto the ear. And it was observable to this purpose that we found a manifest disparity of noise by holding our ears near the sides of the receiver, and near the cover of it: which difference seemed to proceed from that of the texture of the glass, from the structure of the cover (and the cement) through which the sound was propagated from the watch to the ear. But let us prosecute* our experiment. The pump after this being employed, it seemed that from time to time the sound grew fainter and fainter; so that when the receiver was emptied as much as it used to be for the foregoing experiments neither we, nor some strangers that chanced to be then in the room could, by applying our ears to the very sides, hear any noise from within; though we could easily perceive that by the moving of the hand which marked the second minutes, and by that of the balance, that the watch neither stood still nor remarkably varied from its wonted motion. And to satisfy ourselves farther that it was indeed the absence of the air about the watch that hindered us from hearing it, we let in the external air at the stop-cock; and then though we turned the key and stopt the valve yet we could plainly hear the noise made by the balance, though we held our ears sometimes at two foot distance from the outside of the receiver; and this experiment being reiterated into another place succeeded after the like manner. Which seems to prove that whether or no the air be the only, it is at least the principal medium of sounds . . . .

EXPERIMENT XLI.

To satisfy ourselves in some measure about the account upon which respiration is so necessary to the animals that nature hath furnished with lungs, we took (being then unable to procure any other lively bird small enough to be put into the receiver) a lark, one of whose wings had been broken by a shot of a man that we had sent to provide us some birds for our experiment; but notwithstanding this hurt the lark was very lively and did, being put into the receiver, divers times spring up in it to a good height. The vessel being hastily but carefully closed, the pump was diligently plied, and the bird for a while appeared lively enough; but upon a greater exsuction of the air she began manifestly to droop and appear sick, and very soon after was taken with as violent and irregular convulsions as are wont to be observed in poultry when their heads are wrung off. For the bird threw herself over and over two or three times, and died with her breast upward, her head downwards, and her neck awry. And though upon the appearing of these convulsions we turned the stop-cock and let in the air upon her, yet it came too late; whereupon casting our eyes upon one of those accurate dials that go with a pendulum, and were of late ingeniously invented by the noble and learned *Hugenius*,[13] we found that the whole tragedy had been concluded within ten minutes of an hour, part of which time had been employed in cementing the cover to the receiver. Soon after we got a hen-sparrow, which being caught with bird-lime was not at all hurt; when we put her into the receiver, almost to the top of which she would briskly raise herself. The experiment being tried with this bird as it was with the former, she seemed to be dead within seven minutes, one of which were employed in cementing on the cover. But upon the speedy turning of the key, the fresh air flowing in began slowly to revive her, so that after some pantings she opened her eyes and regained her feet, and in about ¼ of an hour after threatened to make an escape at the top of the glass, which had been unstopped to let in the fresh air upon her. But the receiver being closed the second time, she was killed with violent convulsions within five minutes from the beginning of the pumping.[14]

A while after we put in a mouse, newly taken in such a trap as had rather affrighted than hurt him. Whilst he was leaping up very high in the receiver we fastened the cover to it, expecting that an animal used to live in narrow holes with very little fresh air would endure the want of it better than the lately mentioned birds. But though, for a while after the pump was set a work, he continued leaping up as before, yet it was not long ere he began to appear sick and giddy, and to stagger, after which he fell down as dead, but without such violent convulsions as the bird died with. Whereupon, hastily turning the key we let in some fresh air upon him, by which he recovered after a while his senses and his feet, but seemed to continue weak and sick. But at length, growing able to skip as formerly, the pump was plied again for eight minutes, about the middle of which space, if not before, a little air by a mischance got in at the stop cock; and about two minutes after that the mouse divers times leaped up lively enough, though after about two minutes more he fell down quite dead, yet with convulsions far milder than those wherewith the two birds expired. This alacrity so little before his death, and his not dying sooner than at the end of the eighth minute, seemed ascribable to the air (how little soever) that slipt into the receiver. For the first time those convulsions (that if they had not been suddenly remedied had immediately dispatched him) seized on him in six minutes after the pump began to be set a work. These experiments seemed the more strange in regard that during a great part of those few minutes the engine could but inconsiderably rarefy* the air (and that too but by degrees) and at the end of them there remained in the receiver no inconsiderable quantity, as may appear by what we have formerly said of our not being able to draw down water in a tube within much less than a foot of the bottom. With which we likewise considered that by the exsuction of the air and interspersed vapours there was left in the receiver a space some hundreds of times exceeding the bigness of the animal to receive the fuliginous* steams from which exspiration discharges the lungs; and which, in the other cases hitherto known may be suspected, for want of room, to stifle those animals that are closely penned up in too narrow receptacles.

I forgot to mention that having caused these three creatures to be opened* I could, in such small bodies, discover little of what we sought for, and what we might possibly have found in larger animals; for though the lungs of the birds appeared very red, and as it were inflamed, yet that colour being usual enough in the lungs of such winged creatures, deserves not so much our notice as it doth that in almost all the destructive experiments made in our engine the animals appeared to die with violent convulsive motions. From which, whether physicians can gather any thing towards the discovery of the nature of convulsive disorders,* I leave to them to consider.

Having proceeded thus far, though (as we have partly intimated already) there appeared not much cause to doubt but that the death of the forementioned animals proceeded rather from the want of air than that the air was overclogged by the steams of their bodies exquisitely penned up in the glass; yet I, that love not to believe any thing upon conjectures when by a not over-difficult experiment I can try whether it be true or no, thought it the safest way to obviate objections and remove scruples by shutting up another mouse as close as I could in the receiver; wherein it lived about three quarters of an hour, and might possibly have done so much longer had not a Virtuoso of quality,* who in the mean while chanced to make me a visit, desired to see whether or no the mouse could be killed by the exsuction of the ambient air. Whereupon we thought fit to open for a little while an intercourse betwixt the air in the receiver and that without it, that the mouse might thereby (if it were needful for him) be refreshed; and yet we did this without uncementing the cover at the top, that it might not be objected that perhaps the vessel was more closely stopped for the exsuction of the air than before.

The experiment had this event, that after the mouse had lived ten minutes (which we ascribed to this, that the pump for want of having been lately oiled could move but slowly, and could not by him that managed it be made to work as nimbly as it was wont) at the end of that time he died with convulsive fits, wherein he made two or three bounds into the air before he fell down dead.

Nor was I content with this, but for your Lordship's farther satisfaction and my own I caused a mouse that was very hungry to be shut in all night, with a bed of paper for him to rest upon. And to be sure that the receiver was well closed I caused some air to be drawn out of it, whereby, perceiving that there was no sensible leak I presently readmitted the air at the stop-cock, lest the want of it should harm the little animal; and then I caused the engine to be kept all night by the fire-side, to keep him from being destroyed by the immoderate cold of the frosty night. And this care succeeded so well that the next morning I found that the mouse was not only alive but had devoured a good part of the cheese that had been put in with him. And having thus kept him alive full twelve hours or better, we did, by sucking out part of the air, bring him to droop and to appear swelled; and by letting in the air again we soon reduced him to his former liveliness . . . .

I fear your Lordship will now expect that to these experiments I should add my reflections on them, and attempt by their assistance to resolve the difficulties that occur about respiration; since at the beginning I acknowledged a farther enquiry into the nature of that to have been my design in the related trials. But I have yet, because of the inconvenient season of the year, made so few experiments, and have been so little satisfied by those I have been able to make that they have hitherto made respiration appear to me rather a more than a less mysterious thing than it did before. But yet, since they have furnished me with some such new considerations concerning the use of the air as confirms me in my diffidence of the truth of what is commonly believed touching that matter; that I may not appear sullen or lazy, I am content not to decline employing a few hours in setting down my doubts in presenting your Lordship some hints, and in considering whether the trials made in our engine will at least assist us to discover wherein the deficiency lies that needs to be supplied.

And this, my Lord, being all my present design I suppose you will not expect that (as if you knew not or had forgotten what Anatomists are wont to teach) I should entertain you with a needless discourse of the organs of respiration and the variety of their structure in several animals; though if it were necessary and had not been performed by others I should think, with *Galen*,[15] that by treating of the fabrics of living bodies I might compose hymns to the wise author of nature, who, in the excellent contrivance of the lungs and other parts of (those admirable engines) animals, manifests himself to be indeed what the eloquent prophet most justly speaks him, *wonderful in counsel, and excellent in working* ... [Isaiah 28:29].

4 *The Sceptical Chymist* (1661)

The purpose of this, Boyle's best-known book, was to undermine the two leading contemporary theories of the elements out of which compound or "mixt bodies" are composed. For the Aristotelians, still dominant in the "Schools" or universities, all matter could be resolved into air, earth, water, and fire. For the "chymists" (a word which often meant alchemists), especially those influenced by Paracelsus, there were three "spagyrical" (alchemical) or "hypostatical" (substantial) principles, namely salt, sulphur, and mercury. These were not the substances which have these names today, but vague terms, respectively, for the solid, inflammable, and liquid properties of matter. Boyle's attack on these two schools was influential, furthering their declining prestige among the growing scientific community, and was well executed, particularly in its exposure of their reliance on vague and obfuscating language. His own "corpuscular" or "mechanical" theory that the three basic constituents of the universe are matter, motion, and rest (partly derived from Descartes), cautiously outlined in the Appendix here, was articulated more fully in a work on *The Excellency and Grounds of the Mechanical Hypothesis* appended to *The Excellence of Theology* (1674): See *Works*, ed. Birch, IV, pp. 67–78.

The polemical nature of the work accounts for the dialogue form, modelled on Galileo's *Dialogue on the Two Chief World Systems*, complete with an Aristotelian butt, like Galileo's Simplicio.

Source: *The Sceptical Chymist: or Chymico-Physical Doubts and Paradoxes, touching the Experiments whereby Vulgar Spagyrists are wont to endeavour to evince their Salt, Sulphur and Mercury, to be the true Principles of Things*

Robert Boyle (1627–1691)

(London, 1661), as reprinted from the second edition (1679) in *The Works of the Honourable Robert Boyle*, ed. T. Birch, 6 vols. (London, 1772).

[From the *Preface*]

... I observe that of late chymistry begins, as indeed it deserves, to be cultivated by learned men, who before despised it; and to be pretended to by many who never cultivated it, that they may be thought not to be ignorant of it: whence it is come to pass that divers chymical notions about matters philosophical* are taken for granted and employed, and so adopted by very eminent writers, both naturalists* and physicians.* Now this, I fear, may prove somewhat prejudicial to the advancement of solid philosophy: for though I am a great lover of chymical experiments, and though I have no mean esteem of divers chymical remedies,* yet I distinguish these from their notions about the causes of things and their manner of generation. And for aught I can hitherto discern, there are a thousand phænomena in nature, besides a multitude of accidents* relating to the human body, which will scarcely be clearly and satisfactorily made out by them that confine themselves to deduce things from salt, sulphur and mercury, and the other notions peculiar to the chymists,[16] without taking much more notice than they are wont to do of the motions and figures of the small parts of matter, and the other more catholic* and fruitful affections* of bodies. Wherefore it will not, perhaps, be now unseasonable to let our *Carneades*[17] warn men not to subscribe to the grand doctrine of the chymists touching their three hypostatical principles till they have a little examined it, and considered how they can clear it from his objections, divers of which, it is like, they may never have thought on; since a chymist scarce would, and none but a chymist could propose them. I hope also it will not be unacceptable to several ingenious persons, who are unwilling to determine of any important controversy without a previous consideration of what may be said on both sides, and yet have greater desires to understand chymical matters than opportunities of learning them, to find here together, besides several experiments of my own, purposely made to illustrate the doctrine of the elements, divers others scarce to be met with otherwise than scattered among many chymical books; and to find these associated experiments so delivered as that an ordinary reader, if he be but acquainted with the usual chymical terms, may easily enough understand them; and even a wary one may safely rely on them. These things I add because a person any thing versed in the writings of chymists cannot but discern by their obscure, ambiguous, and almost enigmatical way of expressing what

68

they pretend to teach, that they have no mind to be understood at all but by the sons of art* (as they call them), nor to be understood even by these without difficulty and hazardous trials. Insomuch that some of them scarce ever speak so candidly as when they make use of that known chymical sentence; *ubi palam locuti sumus ibi nihil diximus*.[18] And as the obscurity of what some writers deliver makes it very difficult to be understood, so the unfaithfulness of too many others makes it unfit to be relied on. For though unwillingly, yet I must for the truth's sake and the reader's warn him not to be forward to believe chymical experiments when they are set down only by way of prescriptions and not of relations*; that is, unless he that delivers them mentions his doing it upon his own particular knowledge or upon the relations of some credible person, avowing it upon his own experience. For I am troubled, I must complain, that even eminent writers, both physicians and philosophers* whom I can easily name if it be required, have of late suffered themselves to be so far imposed upon as to publish and build upon chymical experiments which questionless they never tried; for if they had they would, as well as I, have found them not to be true ....

And now having said thus much for *Carneades* I hope the reader will give me leave to say something for myself.

And first, if some morose* readers shall find fault with my having made the interlocutors upon occasion compliment with one another, and that I have almost all along written these Dialogues in a style more fashionable than that of meer scholars* is wont to be, I hope I shall be excused by them that shall consider that to keep a due decorum in the discourses, it was fit that in a book written by a gentleman, and wherein only gentlemen are introduced as speakers, the language should be more smooth and the expressions more civil than is usual in the more scholastic way of writing. And indeed I am not sorry to have this opportunity of giving an example how to manage even disputes with civility; whence perhaps some readers will be assisted to discern a difference betwixt bluntness of speech and strength of reason, and find that a man may be a champion for truth without being an enemy to civility; and may confute an opinion without railing at them that holds it; to whom, he that desires to convince and not to provoke them must make some amends by his civility to their persons for his severity to their mistakes; and must say as little else as he can to displease them when he says that they are in an error ....

... And indeed they will much mistake me that shall conclude from what I now publish that I am at defiance with chymistry, or would make my readers so. I hope the *specimina* I have lately published of an attempt to shew the usefulness of chymical experiments[19] to contemplative philosophers will give those that read them other thoughts of me, and I had a design (but wanted opportunity) to publish with these papers an essay I have lying by me, the greater part of which is apologetical for one sort of chymists. And at least, as for those that know me, I hope the pains I have taken in the fire* will both convince them that I am far from being an enemy to the chymists art (though I am no friend to many that disgrace it by professing it) and persuade them to believe me when I declare that I distinguish betwixt those chymists that are either cheats, or but laborants,* and the true *adepti*;* by whom, could I enjoy their conversation,* I would both willingly and thankfully be instructed, especially concerning the nature and generation[20] of metals; and possibly those that know how little I have remitted of my former addictedness to make chymical experiments, will easily believe that one of the chief designs of this sceptical discourse was not so much to discredit chymistry as to give an occasion and a kind of necessity to the more knowing artists* to lay aside a little of their over-great reservedness, and either explicate or prove the chymical theory better than ordinary chymists have done; or, by enriching us with some of their nobler secrets, to evince that their art is able to make amends even for the deficiencies of their theory. And thus much I shall make bold to add, that we shall much undervalue chymistry if we imagine that it cannot teach us things far more useful, not only to physic* but to philosophy, than those that are hitherto known to vulgar chymists. And yet, as for inferior Spagyrists* themselves, they have by their labours deserved so well of the common wealth of learning that methinks it is pity they should ever miss the truth which they have so industriously sought. And though I be no admirer of the theorical part of their art, yet my conjectures will much deceive me if the practical part be not hereafter much more cultivated than hitherto it has been, and do not both employ philosophy and philosophers and hope to make men such . . . .

PHYSIOLOGICAL CONSIDERATIONS

Touching the Experiments wont to be employed to evince either the four Peripatetick[21] Elements or the three chymical Principles of mixt Bodies.

PART OF THE FIRST DIALOGUE.

[*Carneades*, the sceptic, speaks first:]

... Notwithstanding the subtile reasonings I have met with in the books of the Peripateticks, and the pretty experiments that have been shewed me in the laboratories of chymists, I am of so diffident or dull a nature as to think that if neither of them can bring more cogent arguments to evince the truth of their assertion than are wont to be brought, a man may rationally enough retain some doubts concerning the very number of those material ingredients of mixt* bodies, which some would have us call elements, and others principles*. Indeed, when I considered that the tenets concerning the elements are as considerable amongst the doctrines of natural philosophy as the elements themselves are among the bodies of the universe, I expected to find those opinions solidly established upon which so many others are superstructed. But when I took the pains impartially to examine the bodies themselves that are said to result from the blended elements, and to torture* them into a confession of their constituent principles, I was quickly induced to think that the number of the elements has been contended about by philosophers with more earnestness than success ....

[*Eleutherius* takes up Carneades' attack on the scholastics' reliance on syllogism as a chief mode of scientific proof.]

... *Eleutherius* paused not here; but, to prevent* their answer, added almost in the same breath, And I am not a little pleased to find that you are resolved on this occasion to insist rather on experiments than syllogisms. For I, and no doubt you, have long observed that those dialectical subtleties that the schoolmen* too often employ about physiological mysteries are wont much more to declare the wit of him that uses them than increase the knowledge or remove the doubts of sober lovers of truth. And such captious subtleties do indeed often puzzle, and sometimes silence men, but rarely satisfy them; being like the tricks of jugglers, whereby men doubt not but that they are cheated, though oftentimes they cannot declare by what slights they are imposed on. And therefore I think you have done very wisely to make it your business to consider the phænomena relating to the present question which have been afforded by experiments, especially since it might seem injurious to our senses, by whose mediation we acquire so much of the knowledge we have of things corporal, to have recourse to far-fetched and abstracted ratiocinations, to know what are the sensible* ingredients of those sensible things that we daily see and handle, and are supposed to have the liberty to untwist* (if I may so speak) into the primitive bodies they consist of. He annexed* that he wished therefore they would no longer delay his expected satisfaction, if they had not, as he feared they had, forgotten something preparatory to their debate; and that was to lay down what should be all along understood by the word principle or element. *Carneades* thanked him for his admonition, but told him that they had not been unmindful of so requisite a thing. But that being gentlemen, and very far from the litigious humour of loving to wrangle about words or terms or notions as empty, they had, before his coming in, readily agreed promiscuously to use, when they pleased, elements and principles as terms equivalent; and to understand both by the one and the other those primitive and simple bodies of which the mixt ones are said to be composed, and into which they are ultimately resolved . . . .

[*Themistius*, the Aristotelian, defends his school. It has been agreed that the speakers in the dialogue shall not cite texts or authorities in support of their case.]

If you have taken sufficient notice of the late confession which was made by *Carneades*, and which (though his civility dressed it up in complemental expressions) was exacted of him by his justice, I suppose you will be easily made sensible* that I engage in this controversy with great and peculiar disadvantages, besides those which his parts and my personal disabilities would bring to any other cause to be maintained by me against him. For he justly apprehending the force of truth, though speaking by no better a tongue than mine, has made it the chief condition of our duel that I should lay aside the best weapons I have, and those I can best handle; whereas if I were allowed the freedom, in pleading for the four elements, to employ the arguments suggested to me by reason to demonstrate them, I should almost as little doubt of making you a proselyte to those unsevered* teachers, Truth and *Aristotle*, as I do of your candour and your judgment. And I hope you will however consider that that great favourite and interpreter of nature, *Aristotle*, who was (as his *Organum*[22] witnesses) the greatest master of logic that ever lived, disclaimed* the course taken by other petty philosophers (antient and modern) who, not attending the coherence and consequences of their opinions, are more sollicitous to make each particular opinion plausible independently upon the rest, than to frame them all so as not only to be consistent together but to support each other. For that great man, in his vast and comprehensive intellect, so framed each of his notions that, being curiously* adapted into one system, they need not each of them any other defence than that which their mutual coherence gives them; as it is in an arch, where each single stone which, if severed from the rest, would be perhaps defenceless, is sufficiently secured by the solidity and entireness of the whole fabric of which it is a part. How justly this may be applied to the present case, I could easily shew you if I were permitted to declare to you how harmonious *Aristotle*'s doctrine of the elements is with his other principles of philosophy; and how rationally he has deduced their number from that of the combinations of the four first qualities from the kinds of simple motion belonging to simple bodies, and from I know not how many other principles and phænomena of nature, which so conspire with his doctrine of the elements that they mutually strengthen and support each other. But since it is forbidden me to insist on reflections of this kind, I must proceed to tell you that though the assertors of the four elements value reason so highly, and are furnished with arguments enough drawn from thence, to be satisfied that there must be four elements, though no man had ever yet made any sensible trial to discover their number, yet they are not destitute of experience* to satisfy others that are wont to be more swayed by their senses than their reason. And I shall proceed to consider the testimony of experience, when I shall have first advertised you that if men were as perfectly rational as it is to be wished they were, this sensible way of probation* would be as needless as it is wont to be imperfect. For it is much more high and philosophical to discover things *à priori* than *à posteriori*.²³ And therefore the Peripatetics have not been very sollicitous to gather experiments to prove their doctrines, contenting themselves with a few only, to satisfy those that are not capable of a nobler conviction. And indeed they employ experiments rather to illustrate than to demonstrate their doctrines, as astronomers use spheres of pasteboard* to descend to the capacities of such as must be taught by their senses, for want of being arrived to a clear apprehension of purely mathematical notions and truths.

I speak thus *Eleutherius* (adds *Themistius*) only to do right to reason, and not out of diffidence of the experimental proof that I am to alledge. For though I shall name but one, yet it is such a one as will make all others appear as needless as itself will be found satisfactory. For if you but consider a piece of green wood burning in a chimney, you will readily discern in the disbanded* parts of it the four elements, of which we teach it and other mixt bodies to be composed. The fire discovers itself in the flame by its own light; the smoke, by ascending to the top of the chimney and there readily vanishing into air, like a river losing itself in the sea, sufficiently manifests to what element it belongs and gladly returns. The water in its own form, boiling and hissing at the ends of the burning wood, betrays itself to more than one of our senses; and the ashes by their weight, their fieriness, and their dryness, put it past doubt that they belong to the elements of earth. If I spoke (continues *Themistius*) to less knowing persons, I would perhaps make some excuse for building upon such an obvious and easy analysis; but it would be, I fear, injurious not to think such an apology needless to you, who are too judicious either to think it necessary that experiments to prove obvious truths should be far-fetched, or to wonder that among so many mixt bodies that are compounded of the four elements some of them should, upon a slight analysis, manifestly exhibit the ingredients they consist of. Especially since it is very agreeable to the goodness of nature to disclose, even in some of the most obvious experiments that men make, a truth so important and so requisite to be taken notice of by them. Besides, that our analysis, by how much the more obvious we make it, by so much the more suitable it will be to the nature of that doctrine which it is alledged to prove; which being as clear and intelligible to the under-standing as obvious to the sense, it is no marvel the learned part of mankind should so long and so generally imbrace it. For this doctrine is very different from the whimsies of chymists and other modern innovators; of whose hypotheses we may observe, as naturalists do of less perfect animals, that as they are hastily formed so they are commonly short lived. For so these, as they are often framed in one week are, perhaps, thought fit to be laughed at the next; and being built, perchance, but upon two or three experiments are destroyed by a third or fourth. Whereas the doctrine of the four elements was framed by *Aristotle* after he had leisurely considered those theories of former philosophers which are now with great applause revived, as discovered by these later ages, and had so judiciously detected and supplied the errors and defect of former hypotheses concerning the elements that his doctrine of them has been ever since deservedly embraced by the lettered part of mankind: all the philosophers that preceded him having, in their several ages, contributed to the completeness of this doctrine, as those of succeeding times have acquiesced in it. Nor has an hypothesis so deliberately and maturely established been called in question, till in the last century *Paracelsus*[24] and some few other sooty empirics* rather than (as they are fain to call themselves) philosophers, having their eyes darkened and their brains troubled with the smoke of their own furnaces, began to rail at the Peripatetick doctrine, which they were too illiterate to understand, and to tell the credulous world that they could see but three ingredients in mixed bodies; which, to gain themselves the repute of inventors, they endeavoured to disguise by calling them, instead of earth, and fire, and vapour, salt, sulphur, and mercury; to which they gave the canting* title of hypostatical principles. But when they came to describe them they shewed how little they understood what they meant by them by disagreeing as much from one another as from the truth they agreed in opposing: for they deliver their hypotheses as darkly as their processes, and it is almost as impossible for any sober man to find their meaning as it is for them to find their elixir. And indeed, nothing has spread their philosophy but their great brags and undertakings; notwithstanding all which (says *Themistius* smiling) I scarce know any thing they have performed worth wondering at, save that they have been able to draw *Philoponus* to their party, and to engage him to the defence of an unintelligible hypothesis, who knows so well as he does that principles ought to be like diamonds, as well very clear as perfectly solid.

Themistius having after these last words declared by his silence that he had finished his discourse, *Carneades* addressing himself, as his adversary had done, to *Eleutherius*, returned this answer to it: I hoped for a demonstration but I perceive *Themistius* hopes to put me off with an harangue,* wherein he cannot have given me a greater opinion of his parts than he has given me distrust for his hypothesis, since for it even a man of such learning can bring no better arguments. The rhetorical part of his discourse, though it make not the least part of it, I shall say nothing to, designing to examine only the argumentative part, and leaving it to *Philoponus* to answer those passages wherein either *Paracelsus* or chymists are concerned. I shall observe to you that in what he has said besides he makes it his business to do these two things. The one, to propose and make out an experiment to demonstrate the common opinion about the four elements; and the other, to insinuate divers things which he thinks may repair the weakness of his argument from experience, and upon other accounts bring some credit to the otherwise defenceless doctrine he maintains.

To begin then with his experiment of the burning wood, it seems to me to be obnoxious* to not a few considerable exceptions.

And first, if I would now deal rigidly with my adversary I might here make a great question of the very way of probation which he and others employ, without the least scruple, to evince that the bodies commonly called mixt are made up of earth, air, water, and fire, which they are pleased also to call elements; namely, that upon the supposed analysis made by the fire of the former sort of concretes,* there are wont to emerge bodies resembling those which they take for the elements. For, not to anticipate here what I foresee I shall have occasion to insist on when I come to discourse with *Philoponus* concerning the right that fire has to pass for the proper and universal instrument of analyzing mixed bodies; not to anticipate that, I say, if I were disposed to wrangle I might alledge that by *Themistius* his experiment it would appear rather that those he calls elements are made of those he calls mixed bodies, than mixed bodies of the elements. For in *Themistius's* analyzed wood, and in other bodies dissipated* and altered by the fire, it appears, and he confesses, that which he takes for elementary fire and water are made out of the concrete;* but it appears not that the concrete was made up of fire and water. Nor has either he or any man, for aught I know, of his persuasion, yet proved that nothing can be obtained from a body by the fire that was not pre-existent in it . . . .

I consider then (says *Carneades*) in the next place, that there are divers bodies out of which *Themistius* will not prove in haste that there can be so many elements as four extracted by the fire. And I should perchance trouble him, if I should ask him what Peripatetic can shew us (I say not all the four elements, for that would be too rigid a question, but) any one of them extracted out of gold by any degree of fire whatsoever. Nor is gold the only body in nature that would puzzle an Aristotelian (that is no more) to analyze by the fire into elementary bodies; since, for aught I have yet observed, both silver and calcined* Venetian talc, and some other concretes not necessary here to be named, are so fixt that to reduce any of them into four heterogeneous substances has hitherto proved a task much too hard not only for the disciples of *Aristotle* but those of *Vulcan*,* at least whilst the latter have employed only fire to make the analysis.

The next argument (continues *Carneades*) that I shall urge against *Themistius's* opinion shall be this: that as there are divers bodies whose analysis by fire cannot reduce them into so many hetero-geneous substances or ingredients as four, so there are others which may be reduced into more, as the blood (and divers other parts) of men and other animals; which yield when analyzed five distinct substances, phlegm,* spirit, oil, salt and earth, as experience has shewn us in distilling man's blood, hartshorn,* and divers other bodies that, belonging to the animal-kingdom, abound with a not uneasily sequestrable* salt . . . .

PART III.

[*Carneades* continues:]

Wherefore I will now proceed to my third general consideration, which is that it does not appear that three is precisely and universally the number of the distinct substances or elements whereunto mixt bodies are resoluble by the fire. I mean, that it is not proved by chymists that all the compound bodies which are granted to be perfectly mixt are upon their chymical analysis divisible each of them into just three distinct substances, neither more nor less, which are wont to be looked upon as elementary, or may as well be reputed so as those that are so reputed.

To talk then again according to such principles as I then made use of, I shall represent* that if it be granted rational to suppose, as I then did, that the elements consisted at first of certain small and primary coalitions* of the minute particles of matter into corpuscles very numerous, and very like each other, it will not be absurd to conceive that such primary clusters may be of far more sorts than three or five; and consequently, that we need not suppose that in each of the compound bodies we are treating of there should be found just three sorts of such primitive coalitions as we are speaking of.

And if, according to this notion, we allow a considerable number of differing elements, I may add that it seems very possible that to the constitution of one sort of mixt bodies two kinds of elementary ones may suffice (as I lately exemplified to you in that most durable concrete, glass,) another sort of mixts may be composed of three elements, another of four, another of five, and another perhaps of many more. So that, according to this notion, there can be no determinate number assigned as that of the elements of all sorts of compound bodies whatsoever; it being very probable that some concretes consist of fewer, some of more elements. Nay it does not seem impossible, according to these principles, but that there may be two sorts of mixts, whereof the one may not have any of all the same elements as the other consists of: as we oftentimes see two words, whereof one has not any one of the letters to be met with in the other; or as we often meet with divers electuaries* in which no ingredient (except sugar) is common to any two of them. I will not here debate whether there may not be a multitude of these corpuscles, which by reason of their being primary and simple might be called elementary, if several sorts of them should convene to compose any body which are as yet free, and neither as yet contexed* and entangled with primary corpuscles of other kinds, but remain liable to be subdued and fashioned by seminal* principles, or the like powerful and transmuting agent; by whom they may be so connected among themselves, or with the parts of the bodies, as to make the compound bodies whose ingredients they are resoluble into more, or other elements than those that chymists have hitherto taken notice of . . . .

PART IV.

And thus much (says *Carneades*) may suffice to be said of the number of the distinct substances separable from mixt bodies by the fire. Wherefore I now proceed to consider the nature of them, and shew you that though they seem homogeneous bodies, yet have they not the purity and simplicity that is requisite to elements; and I should immediately proceed to the proof of my assertion, but that the confidence wherewith chymists are wont to call each of the substances we speak of by the name of sulphur or mercury, or the other of the hypostatical principles, and the intolerable ambiguity they allow themselves in their writings and expressions makes it necessary for me, in order to the keeping you either from mistaking me or thinking I mistake the controversy, to take notice to you and complain of the unreasonable liberty they give themselves of playing with names at pleasure. And indeed if I were obliged in this dispute to have such regard to the phraseology of each particular chymist, as not to write any thing which this or that author may not pretend, not to contradict this or that sense which he may give us as occasion serves to his ambiguous expressions, I should scarce know how to dispute nor which way to turn myself. For I find that even eminent writers (such as *Raymund Lully*,[25] *Paracelsus*, and others) do so abuse the terms they employ that as they will now and then give divers things one name, so they will oftentimes give one thing many names; and some of them (perhaps) such as do much more properly signify some distinct body of another kind. Nay, even in technical words or terms of art they refrain not from this confounding* liberty; but will, as I have observed, call the same substance sometimes the sulphur, and sometimes the mercury of a body. And now I speak of mercury, I cannot but take notice that the descriptions they give us of that principle or ingredient of mixt bodies are so intricate that even those that have endeavoured to polish* and illustrate* the notions of the chymists, are fain to confess that they know not what to make of it either by ingenuous acknowledgments, or descriptions that are not intelligible.

I must confess (says *Eleutherius*) I have in the reading of *Paracelsus* and other chymical authors been troubled to find that such hard* words and equivocal expressions as you justly complain of do, even when they treat of principles, seem to be studiously affected by those writers, whether to make themselves to be admired by their readers and their art appear more venerable and mysterious, or (as they would have us think) to conceal from them a knowledge themselves judge inestimable.

But whatever (says *Carneades*) these men may promise themselves from a canting way of delivering the principles of nature, they will find the major part of knowing men so vain as when they understand not what they read, to conclude that it is rather the writer's fault than their own. And those that are so ambitious to be admired by the vulgar that rather than go without the admiration of the ignorant they will expose themselves to the contempt of the learned, those shall, by my consent, freely enjoy their option. As for the mystical writers scrupling to communicate their knowledge, they might less to their own disparagement and to the trouble of their readers have concealed it by writing no books than by writing bad ones. If *Themistius* were here he would not stick to say that chymists write thus darkly, not because they think their notions too precious to be explained, but because they fear that if they were explained men would discern that they are far from being precious. And indeed I fear that the chief reason why chymists have written so obscurely of their three principles may be, that not having clear and distinct notions of them themselves they cannot write otherwise than confusedly of what they but confusedly apprehend: not to say that divers of them, being conscious to the invalidity of their doctrine, might well enough discern that they could scarce keep themselves from being confuted but by keeping themselves from being clearly understood. But though much may be said to excuse the chymists when they write darkly and ænigmatically about the preparation of their elixir* and some few other grand arcana,* the divulging of which they may, upon grounds plausible enough, esteem unfit; yet when they pretend to reach the general principles of natural philosophers this equivocal way of writing is not to be endured. For in such speculative enquiries, where the naked knowledge of the truth is the thing principally aimed at, what does he teach me worth thanks that does not, if he can, make his notion intelligible to me, but by mystical terms and ambiguous phrases darkens what he should clear up, and makes me add the trouble of guessing at the sense of what he equivocally expresses to that of examining the truth of what he seems to deliver? And if the matter of the philosopher's stone,* and the manner of preparing it, be such mysteries as they would have the world believe them, they may write intelligibly and clearly of the principles of mixt bodies in general without discovering* what they call the great work.* But for my part (continues *Carneades*) what my indignation at this un-philosophical way of teaching principles has now extorted from me is meant chiefly to excuse myself if I shall hereafter oppose any particular opinion or assertion that some follower of *Paracelsus* or any eminent artist* may pretend not to be his master's. For as I told you long since, I am not obliged to examine private men's writings (which were a labour as endless as unprofitable), being only engaged to examine those opinions about the *tria prima** which I find those chymists I have met with to agree in most; and I doubt not but my arguments against their doctrine will be in great part easily enough applicable even to those private opinions which they do not so directly and expressly oppose. And indeed, that which I am now entering upon, being the consideration of the things themselves whereinto Spagyrists resolve mixt bodies by the fire, if I can shew that these are not of an elementary nature it will be no great matter what names these or those chymists have been pleased to give them. And I question not that to a wise man, and consequently to *Eleutherius*, it will be less considerable to know what men have thought of things than what they should have thought.

In the fourth and last place, then, I consider that as generally as chymists are wont to appeal to experience, and as confidently as they use to instance the several substances separated by the fire from a mixt body as a sufficient proof of their being its component elements; yet those differing substances are many of them far enough from elementary simplicity, and may be yet looked upon as mixt bodies, most of them also retaining somewhat at least, if not very much, of the nature of those concretes whence they were forced. I am glad (says *Eleutherius*) to see the vanity or envy of the canting chymists thus discovered and chastised; and I could wish that learned men would conspire together to make these deluding writers sensible that they must no longer hope with impunity to abuse the world. For whilst such men are quietly permitted to publish books with promising titles, and therein to assert what they please, and contradict others and even themselves as they please, with as little danger of being confuted as of being understood, they are encouraged to get themselves a name at the cost of the readers, by finding that intelligent men are wont, for the reason newly mentioned, to let their books and them alone, and the ignorant and credulous (of which the number is still much greater than that of the other) are forward to admire most what they least understand. But if judicious men, skilled in chymical affairs,

shall once agree to write clearly and plainly of them and thereby keep
men from being stunned, as it were, or imposed upon by dark or
empty words; it is to be hoped that these men, finding that they can
no longer write impertinently and absurdly without being laughed at
for doing so, will be reduced either to write nothing, or books that
may teach us something and not rob men, as formerly, of invaluable
time; and so ceasing to trouble the world with riddles or impertinen-
cies, we shall either by their books receive an advantage or by their
silence escape an inconvenience.

But after all this is said (continues *Eleutherius*) it may be represented in favour of the chymists that, in one regard the liberty they
take in using names, if it be excusable at any time, may be more so
when they speak of the substances whereinto their analysis resolves
mixt bodies: since as parents have the right to name their own
children, it has ever been allowed to the authors of new inventions to
impose names upon them. And therefore the subjects we speak of
being so the productions of the chymists art as not to be otherwise
but by it obtainable, it seems but equitable to give the artists leave to
name them as they please; considering also that none are so fit and
likely to teach us what those bodies are as they to whom we owed
them.

I told you already (says *Carneades*) that there is great difference
betwixt the being able to make experiments and the being able to give
a philosophical account of them. And I will not now add that many a
mine-digger may meet, whilst he follows his work, with a gem or a
mineral which he knows not what to make of till he shews it a jeweller
or a mineralist to be informed what it is. But that which I would
rather have here observed is that the chymists I am now in debate
with have given up the liberty you challenged for them of using
names at pleasure, and confined themselves by their descriptions,
though but such as they are, of their principles. So that although they
might freely have called any thing their analysis presents them with,
either sulphur, or mercury, or gas*, or blas,* or what they pleased, yet
when they have told me that sulphur (for instance) is a primogeneal*
and simple body, inflammable, odorous, &c., they must give me leave
to disbelieve them if they tell me that a body that is either compounded or uninflammable is such a sulphur[26], and to think they play with
words when they teach that gold and some other minerals abound
with an incombustible sulphur, which is as proper an expression as
a sunshine night, or a fluid ice . . . .

Robert Boyle (1627–1691)

PART V

... And from what has been hitherto deduced (continues *Carneades*) we may learn what to judge of the common practice of those chymists who, because they have found that divers compound bodies (for it will not hold in all) can be resolved into, or rather can be brought to afford two or three differing substances more than the soot and ashes whereunto the naked fire commonly divides them in our chimneys, cry up their own sect for the invention of a new philosophy; some of them, as *Helmont*,[27] &c., styling themselves philosophers by the fire; and the most part not only ascribing but, as far as in them lies, engrossing to those of their sect the title of Philosophers.

But alas, how narrow is this philosophy, that reaches but to some of those compound bodies which we find but upon or in the crust or outside of our terrestrial globe, which is itself but a point in comparison of the vast extended universe, of whose other and greater parts the doctrine of the *tria prima* does not give us an account! For what does it teach us, either of the nature of the sun, which astronomers affirm to be eight score and odd times bigger than the whole earth? or of that of those numerous fixt stars which, for aught we know, would very few, if any of them, appear inferior in bulk and brightness to the sun if they were as near us as he? What does the knowing that salt, sulphur and mercury are the principles of mixt bodies inform us of the nature of that vast, fluid, and ætherial substance that seems to make up the interstellar and consequently much the greatest part of the world? For as the opinion commonly ascribed to *Paracelsus*, as if he would have not the only four Peripatetick elements but even the celestial parts of the universe to conflict of his three principles, since the modern chymists themselves have not thought so groundless a conceit worth their owning I shall not think it worth my confuting.

But I should perchance forgive the hypothesis I have been all this while examining if, though it reaches but to a very little part of the world, it did at least give us a satisfactory account of those things to which it is said to reach. But I find not that it gives us any other than a very imperfect information even about mixt bodies themselves: for how will the knowledge of the *tria prima* discover to us the reason why the load-stone* draws a needle and disposes it to respect the poles, and yet seldom precisely points at them? How will this hypothesis teach us how a chick is formed in the egg, or how the seminal principles of mint, pompions,* and other vegetables that I mentioned to you above, can fashion water into various plants each of them endowed with its peculiar and determinate shape and with divers specifick and discriminating qualities? How does this hypothesis shew us how much salt, how much sulphur, and how much mercury must be taken to make a chick or a pompion? And if we know that, what principle is it that manages these ingredients, and contrives (for instance) such liquors* as the white and yolk of an egg into such a variety of textures as is requisite to fashion the bones, veins, arteries, nerves, tendons, feathers, blood, and other parts of a chick? and not only to fashion each limb, but to connect them all together after that manner that is most congruous to the perfection of the animal which is to consist of them? For to say that some more fine and subtile part of either or all the hypostatical principles is the director in all this business and the architect of all this elaborate structure, is to give one occasion to demand again what proportion and way of mixture of the *tria prima* afforded this architectonick* spirit, and what agent made so skilful and happy a mixture? And the answer to this question, if the chymists will keep themselves within their three principles, will be liable to the same inconvenience that the answer to the former was. And if it were not to intrench upon the theme of a friend of ours here present, I could easily prosecute* the imperfections of the vulgar chymists philosophy and shew you that by going about to explicate by their three principles, I say not all the abstruse properties of mixt bodies, but even such obvious and more familiar phænomena as fluidity and firmness, the colours and figures* of stones, minerals, and other compound bodies, the nutrition of either plants or animals, the gravity of gold or quicksilver compared with wine or spirit of wine; by attempting, I say, to render a question of these (to omit a thousand others as difficult to account for) from any proportion of the three simple ingredients, chymists will be much more likely to discredit themselves and their hypothesis than satisfy an intelligent inquirer after truth ....

... I acknowledge the great service that the labours of chymists have done the lovers of useful learning; nor even, on this occasion, shall their arrogance hinder my gratitude. But since we are as well examining the truth of their doctrine as the merit of their industry, I must, in order to the investigation of the first, continue a reply to talk at the rate of* the part I have assumed; and tell you that when I acknowledge the usefulness of the labours of Spagyrists to natural philosophy I do it upon the score of their experiments, not upon that of their speculations; for it seems to me that their writings, as their furnaces, afford as well smoke as light, and do little less obscure some subjects than they illustrate others. And though I am unwilling to deny that it is difficult for a man to be an accomplished naturalist that is a stranger to chymistry, yet I look upon the common operations and practices of chymists almost as I do on the letters of the alphabet, without whose knowledge it is very hard for a man to become a philosopher; and yet that knowledge is very far from being sufficient to make him one . . . .

A Paradoxical Appendix to the foregoing Treatise

PART VI.

[*Eleutherius* challenges *Carneades* to take his scepticism to its logical conclusion by questioning "whether there be any elements at all". *Carneades* demurs at having to "maintain so great and so invidious a paradox",[28] but does so.]

And to prevent mistakes, I must advertise you that I now mean by elements, as those chymists that speak plainest do by their principles, certain primitive and simple, or perfectly unmingled bodies; which not being made of any other bodies or of one another, are the ingredients of which all those perfectly mixt bodies are immediately compounded, and into which they are ultimately resolved. Now whether there be any one such body to be constantly met with in all and each of those that are said to be elemented bodies, is the thing I now question.

By this state of the controversy you will, I suppose, guess that I need not be so absurd as to deny that there are such bodies as earth and water, and quicksilver and sulphur. But I look upon earth and water as component parts of the universe, or rather of the terrestrial globe, not of all mixt bodies. And though I will not peremptorily deny that there may sometimes either a running* mercury or a combustible substance be obtained from a mineral, or even a metal, yet I need not concede either of them to be an element in the sense above declared, as I shall have occasion to shew you by and by . . . .

I should tell you that I have sometimes thought it not unfit that to the principles which may be assigned to things as the world is now constituted we should, if we consider the great mass of matter as it was whilst the universe was in making, add another, which may conveniently enough be called an architectonic principle or power. By which I mean those various determinations, and that skilful guidance of the motions of the small parts of the universal matter by the most wise author of things, which were necessary at the beginning to turn that confused chaos into this orderly and beautiful world; and especially to contrive the bodies of animals and plants, and the seeds of those things whose kinds were to be propagated. For I confess I cannot well conceive how from matter, barely put into motion and then left to itself, there could emerge such curious fabricks as the bodies of men and perfect animals, and such yet more admirably contrived parcels* of matter as the seeds* of living creatures.

I should likewise tell you upon what grounds, and in what sense, I suspected the principles of the world as it now is to be three, matter, motion and rest. I say as the world now is, because the present fabric of the universe, and especially the seeds* of things, together with the established course of nature, is a requisite or condition upon whose account divers things may be made out by our three principles, which otherwise would be very hard, if possible, to explicate.

I should moreover declare in general (for I pretend not to be able to do it otherwise) not only why I conceive that colours, odours, tastes, fluidness and solidity, and those other qualities that diversify and denominate bodies, may intelligibly be deduced from these three, but how two of the three Epicurean[29] principles (which, I need not tell you, are magnitude, figure, and weight) are themselves deducible from matter and motion; since the latter of these variously agitating, and, as it were, distracting the former, must needs disjoin its parts, which being actually separated must each of them necessarily both be of some size, and obtain some shape or other. Nor did I add to our principles the Aristotelian privation,* partly for other reasons which I must not now stay to insist on; and partly because it seems to be rather an antecedent or a *terminus à quo** than a true principle, as the starting-post is none of the horse's legs or limbs.

I should also explain why and how I made rest to be, though not so considerable a principle of things as motion, yet a principle of them; partly because it is (for aught we know) as antient at least as it, and depends not upon motion nor any other quality of matter; and partly because it may enable the body in which it happens to be, both to continue in a state of rest till some external force put it out of that state, and to concur to the productions of divers changes in the bodies that hit against it, by either quite stopping or lessening their motion (whilst the body formerly at rest receives all or part of it into itself) or else by giving a new biass, or some other modification to motion, that is, to the grand and primary instrument whereby nature produces all the changes and other qualities that are to be met with in the world.

I should likewise, after all this, explain to you how, although matter, motion, and rest seemed to me to be the catholic* principles of the universe, I thought the principles of particular bodies might be commodiously enough reduced to two, namely matter and (what comprehends the two other, and their effects) the result, or aggregate, or complex of those accidents, which are the motion or rest (for in some bodies both are not to be found), the bigness, figure,* texture, and the thence resulting qualities of the small parts which are necessary to entitle the body whereto they belong to this or that peculiar denomination; and discriminating it from others, to appropriate it to a determinate kind of things (as yellowness, fixedness, such a degree of weight, and of ductility,* do make the portion of matter wherein they concur to be reckoned among perfect metals and obtain the name of gold). This aggregate or result of accidents you may, if you please, call either structure or texture (though indeed that do not so properly comprehend the motion of the constituent parts, especially in case some of them be fluid), or what other appellation shall appear most expressive. Or if, retaining the vulgar term, you will call it the form of the thing it denominates, I shall not much oppose it provided the word be interpreted to mean but what I have expressed, and not a scholastic substantial form,[30] which so many intelligent men profess to be to them altogether unintelligible.

But (says *Carneades*) if you remember that it is a sceptic speaks to you, and that it is not so much my present task to make assertions as to suggest doubts, I hope you will look upon what I have proposed rather as a narrative of my former conjectures touching the principles of things than as a resolute declaration of my present opinions of them; especially since although they cannot but appear very much to their disadvantage if you consider them as they are proposed, without those reasons and explanations by which I could perhaps make them appear much less extravagant.

THE CONCLUSION.

... I think I may presume that what I have hitherto discoursed will induce you to think that chymists have been much more happy in finding experiments than the causes of them, or in assigning the principles by which they may best be explained. And indeed, when in the writings of *Paracelsus* I meet with such phantastic and unintelligible discourses as that writer often puzzles and tires his reader with, fathered upon such excellent experiments, as though he seldom clearly teaches I often find he knew; methinks the chymists, in their searches after truth, are not unlike the navigators of *Solomon's Tarshish* fleet, who brought home from their long and tedious voyages not only gold, and silver, and ivory, but apes and peacocks too.[31] For so the writings of several (for I say not all) of your hermetic philosophers present us, together with divers substantial and noble experiments, theories, which either like peacocks' feathers make a great shew but are neither solid nor useful; or else like apes, if they have some appearance of being rational are blemished with some absurdity or other that, when they are attentively considered, make them appear ridiculous.

Henry Power

(1623–1668)

5 Experimental Philosophy (1664)

Power qualified as a doctor at Cambridge, where he became friendly with Sir Thomas Browne. He practised medicine in Halifax, Yorkshire, but continued to pursue his scientific interests, being admitted to the Royal Society in 1661. His single book also contains experiments on atmospheric pressure, in verification of Boyle's work, and on magnetism.

Source: *Experimental Philosophy, In Three Books: Containing New Experiments, Microscopical, Mercurial, Magnetical. With some Deductions, and Probable Hypotheses, raised from them, in Avouchment and Illustration of the now famous Atomical Hypothesis* (London, 1664). The facsimile reprint edited by Marie Boas Hall (New York and London, 1966) includes a valuable introduction and an appendix containing Power's own corrections and additional observations (pp. 197–207).

[from the *Preface*]

...."The knowledge of Man" (saith the learn'd *Verulam*) "hath hitherto been determin'd by the view or sight, so that whatsoever is invisible, either in respect of the fineness of the Body it self or the smalness of the parts, or of the subtilty of its motion, is little enquired; and yet these be the things that govern Nature principally."[1] How much therefore are we oblig'd to modern Industry,* that of late hath discover'd this advantageous Artifice of Glasses,* and furnish'd our necessities with such artificial Eys that now neither the fineness of the Body, nor the smalness of the parts, nor the subtilty of its motion, can secure them from our discovery? And indeed, if the Dioptricks* further prevail, and that darling Art could but perform what the Theorists in Conical sections* demonstrate, we might hope, ere long, to see the Magnetical Effluviums* of the Loadstone, the Solary Atoms of light (or *globuli ætherei* of the renowned *Des-Cartes*[2]), the springy* particles of Air, the constant and tumultuary motion of the Atoms of all fluid Bodies, and those infinite, insensible* Corpuscles (which daily produce those prodigious (though common) effects amongst us). And though these hopes be vastly hyperbolical, yet who can tel how far Mechanical Industry* may prevail; for the process of

Art is indefinite, and who can set a *non-ultra*³ to her endevours? I am sure, if we look backwards at what the Dioptriks hath already perform'd, we cannot but conclude such Prognosticks to be within the circle of possibilities, and perhaps not out of the reach of futurity to exhibit. However, this I am sure of, That without some such Mechanical assistance our best Philosophers will but prove empty Conjecturalists, and their profoundest Speculations herein but gloss'd outside* Fallacies; like our Stage-scenes, or Perspectives, that shew things inwards, when they are but superficial paintings.

For, to conclude with that doubly Honourable (both for his parts and parentage) *Mr. Boyle*, When a Writer, saith he, acquaints me onely with his own thoughts or conjectures, without inriching his discourse with any real Experiment or Observation, if he be mistaken in his Ratiotination,* I am in some danger of erring with him, and at least am like to lose my time, without receiving any valuable compensation for so great a loss. But if a Writer endevours, by delivering new and real Observations or Experiments, to credit his Opinions, the Case is much otherwayes; for, let his Opinions be never so false (his Experiments being true) I am not oblig'd to believe the former, and am left at my liberty to benefit my self by the latter. And though he have erroneously superstructed upon his Experiments, yet the Foundation being solid, a more wary Builder may be very much further'd by it, in the erection of a more judicious and consistent Fabrick*.⁴

MICROSCOPICAL OBSERVATIONS.

OBSERVAT. I. OF THE FLEA.

It seems as big as a little Prawn or Shrimp, with a small head, but in it two fair eyes globular and prominent of the circumference of a spangle;* in the midst of which you might (through the diaphanous Cornea) see a round blackish spot, which is the pupil or apple of the eye, beset round with a greenish glittering circle, which is the Iris, (as vibrissant* and glorious as a Cats eye) most admirable to behold.

How critical* is Nature in all her works! that to so small and contemptible an Animal hath given such an exquisite fabrick of the eye, even to the distinction of parts. He has also a very long neck, jemmar'd* like the tail of a Lobstar, which he could nimbly move any way. His head, body, and limbs also, be all of blackish armour-work, shining and polished with jemmar's, most excellently contrived for the nimble motion of all the parts: nature having armed him thus *Cap-a-pe** like a Curiazier* in warr, that he might not be hurt by the great leaps he takes. To which purpose also he hath so excellent an eye, the better to look before he leap: to which add this advantageous contrivance of the joynts of his hinder legs which bend backwards towards his belly, and the knees or flexure of his fore-legs forwards (as in most quadrupeds) that he might thereby take a better rise when he leaps. His feet are slit into claws or talons, that he might the better stick to what he lights upon. He hath also two pointers before which grow out of the forehead, by which he tryes and feels all objects, whether they be edible or no. His neck, body, and limbs are also all beset with hairs and bristles, like so many Turnpikes,* as if his armour was palysado'd* about by them. At his snout is fixed a Proboscis, or hollow trunk or probe, by which he both punches the skin, and sucks the blood through it, leaving that central spot in the middle of the Flea-biting, where the probe entered.

One would wonder at the great strength lodged in so small a Receptacle, and that he is not able onely to carry his whole armour about him, but will frisk and curvet so nimbly with it. Stick a large brass pin through his tayl and he will readily drag it away. I have seen a chain of gold (at *Tredescants*[5] famous reconditory* of Novelties) of three hundred links, though not above an inch long, both fastned to, and drawn away by a Flea ... Yea, we have heard it credibly reported, saith [*Muffet*],[6] that a Flea hath not onely drawn a gold Chain, but a golden Charriot also with all its harness and accoutrements fixed to it, which did excellently set forth the Artifice of the Maker, and Strength of the Drawer; so great is the mechanick power which Providence has immur'd within these living walls of Jet.

OBSERVAT. III. THE COMMON FLY.

It is a very pleasant Insect to behold. Her body is as it were from head to tayl studded with silver and black Armour, stuck all over with great black Bristles, like Porcupine quills, set all in parallel order, with their ends pointing all towards the tayl; her wings look like a Sea-fan with black thick ribs or fibers, dispers'd and branch'd through them, which are webb'd between with a thin membrane or film, like a slice of Muscovy-glasse.* She hath a small head which she can move or turn any way. She hath six legs, but goes onely but upon four; the two foremost she makes use of instead of hands, with which you may often see her wipe her mouth and nose, and take up any thing to eat. The other four legs are cloven and arm'd with little clea's* or tallons* (like a Catamount*) by which she layes hold on the rugosities* and asperities* of all bodies she walks over, even to the supportance of her self, though with her back downwards and perpendicularly invers'd to the Horizon. To which purpose also the wisdom of Nature hath endued her with another singular Artifice, and that is a fuzzy kinde of substance like little sponges, with which she hath lined the soles of her feet, which substance is always repleated* with a whitish viscous liquor, which she can at pleasure squeeze out, and so sodder* and be-glew her self to the plain she walks on, which otherways her gravity would hinder (were it not for this contrivance) especially when she walks in those inverted positions.

But of all things her eyes are most remarkable, being exceeding large, ovally protuberant and most neatly dimpled with innumerable little cavities like a small grater* or thimble, through which seeming perforations you may see a faint reddish colour (which is the blood in the eyes, for if you prick a pin through the eye, you shall finde more blood there, than in all the rest of her body.) The like foraminulous* perforations or trelliced* eyes are in all Flyes, more conspicuously in Carnivorous or Flesh-Flyes, in the Stercorary* or Yellow Flyes that feed upon Cow-dung. The like eyes I have also found in divers other Insects ....

OBSERVAT. IV. THE GRAY, OR HORSE-FLY.

Her eye is an incomparable pleasant spectacle: 'tis of a semisphaeroidal figure, black and waved,* or rather indented all over with a pure Emerauld-green, so that it looks like green silk Irish-stich,* drawn upon a black ground,* and all latticed* or chequered with dimples like Common Flyes, which makes the Indentures* look more pleasantly. Her body looks like silver in frost-work, onely fring'd all over with white silk. Her legs all joynted and knotted like the plant call'd *Equisetum* or Horse-tayle, and all hairy and slit at the ends into two toes, both which are lined with two white sponges or fuzballs as is pre-observ'd in Common Flyes. After her head is cut off, you shall most fairly* see (just at the setting on of her neck) a pulsing particle* (which certainly is the heart) to beat for half an hour most orderly and neatly through the skin.

OBSERVAT. XII. MITES IN CHEESE.

They appeared some bigger, some less; the biggest appeared equal to a Nutmeg; in shape they seem'd oval and obtus'd* towards the tail. Their colour resembled that of Mother of pearl, or Common pearl, and reflected the light of the Sun in some one point, according to their various positions, as pearl doth: so that it seems they are sheath'd* and crustaceous* Animals (as Scarabees and such like Insects are.) I could perfectly see the divisions of the head, neck, and body. To the small end of the oval Body was fastned the head, very little in proportion to the body, its mouth like that of a Mole, which it open'd and shutt; when open'd, it appear'd red within. The eyes also, like two little dark spots, are discernable. Near to the head were four legs fastned, two on each side; the legs were just like to those in a Louse, Jemmar'd* and Transparent. She has two little pointers at the snout; nay, you may see them sometimes, if you happily take the advantage, like so many *Ginny-Pigs*, munching and chewing the cud. About the head and tail are stuck long hairs or bristles. Some we could see (as little, even in the Glass,* as a Mustard-seed) yet perfectly shap'd and organiz'd. We also saw divers Atoms somewhat Transparent like eggs, both in form and figure. Nay, in these moving Atoms, I could not onely see the long bristles formerly specified, but also the very hairs which grew out of their leggs, which leggs themselves are smaller than the smallest hair our naked eyes can discover. What rare Considerations might an Ingenious Speculator take up here, even from this singular Experiment of the strange and most prodigious skilfulness of Nature in the fabrick of so Minute an Animal (a thousand whereof do not weigh one single grain, for one seed of Tobacco is bigger than any of them) and yet how many thousand parts of Matter must go to make up this heterogeneous Contexture? For, besides the parts inservient* to Nutrition, Sensation, and Motion, how small and thin must the liquors be that circulate through the pipes and vessels disseminated through those parts? nay, How incomprehensibly subtil must the Animal-spirits* be, that run to and fro in Nerves included in such prodigiously little spindle-shank'd leggs?

OBSERVAT. XXI. COMMON GRASSHOPPERS.

In those Common Grasshoppers, both great and little, which are so frequent at hay-time with us, there are some things remarkable. First, Their Eyes, which like other Insects are foraminulous; nay, we have taken the Cornea or outward Film of the Eye quite off, and clensed it so from all the pulpous matter which lay within it, that it was clear and diaphanous like a thin film of Sliffe* or Muscovy-glass, and then looking again on it in the *Microscope*, I could plainly see it foraminulous as before.

You shall in all Grasshoppers see a green Film or Plate (like a Corslet) which goes over the neck and shoulders, which if you lift up with a pin, you may see their heart play, and beat very orderly for a long time together.

The like curious Lattice-work I have also observ'd in the crustaceous *Cornea* of the Creckets Eye, which I have carefully separated from all the matter which stuff'd it within, which certainly is their Brain; as hereafter shall be made more probable.

OBSERVAT. XXII. THE ANT, EMMET OR PISMIRE.

This little Animal is that great Pattern of Industry and Frugality: To this Schoolmaster did *Solomon*[7] send his Sluggard, who in those virtues not onely excels all Insects, but most men. Other excellent Observables there are in so small a fabrick: As the *Herculean* strength of its body, that it is able to carry its triple weight and bulk: The Agility of its limbs, that it runs so swiftly: The equality of its Motion, that it trips so nimbly away without any saliency or leaping, without any fits or starts in its Progression. Her head is large and globular, with a prominent Snout: her eye is of a very fair black colour, round, globular, and prominent, of the bigness of a Pea, foraminulous and latticed like that of other Insects: her mouth (in which you may see something to move) is arm'd with a pair of pincers, which move laterally, and are indented on the inside like a Saw, by which she bites, and better holds her prey; and you may often see them carry their white oblong eggs in them for better security.

OBSERVAT. XLVII. A NITT.

A Nitt is an Egge glewed by some viscous matter to the sides of the hair it sticks to; it is Oval in shape, white in colour, and full of transparent Liquor or Gelly, and seems to be cased in a brittle Shell by the crackling it makes 'twixt your nails. In the same manner appears a Nitt in a Horse's hair: *Muffet* will needs have it a quick, or rudely-shaped Animal. Thus discursive Argumentation and Rational probabilities mislead men in the Wilderness of Enquiry; but he that travels by the Clew* which his own sense and ocular observation has spun out, is likeliest to trace the securest path, and go furthest into the Maze and Labyrinth of Truth.

Some Considerations, Corollaries, and Deductions, Anatomical, Physical, and Optical, drawn from the former Experiments and Observations.

First, therefore, it is Ocularly manifest from the former Observations, that, as perfect Animals have an incessant motion of their Heart, and Circulation of their Bloud (first discovered by the illustrious Doctor *Harvey*,)[8] so in these puny *automata**, and exsanguineous* pieces of

Nature, there is the same pulsing Organ, and Circulation of their Nutritive Humour* also: as is demonstrated by OBSERV. fourth, sixth, seventeenth, &c.

Nay, by OBSERV. sixth, it is plain that a Louse is a Sanguineous Animal, and hath both an Heart and Auricles, the one manifestly preceding the pulse of the other; and hath a purple Liquor or Bloud, which circulates in her (as the Noblest sort of Animals have) which though it be onely conspicuous in its greatest bulk at the heart, yet certainly it is carried up and down in Circulatory Vessels; which Veins and Arteries are so exceeding little, that both they and their Liquor are insensible. For certainly, if we can at a Lamp-Furnace draw out such small Capillary Pipes of Glass that the reddest Liquor in the World shall not be seen in them (which I have often tried and done;) how much more curiously can Nature weave the Vessels of the Body; nay, and bore them too with such a Drill, as the Art of man cannot excogitate. Besides, we see, even in our own Eyes, that the Sanguineous Vessels that run along the white of the eye (nay and probably into the diaphanous humours also) are not discernable, but when they are preter-naturally distended in an Ophthalmia,* and so grow turgent* and conspicuous.

THE FIFTH COROLLARY. ANATOMICAL CONSIDERATIONS ABOUT THE EYE.

Our next Reflections shall be made upon the Eye, to admire as well as contemplate Nature's variety in the constructure and conformation of so excellent an Organ. The two Luminaries of our *Microcosm*, which see all other things, cannot see themselves, nor discover the excellencies of their own Fabrick. Nature, that excellent Mistress of the Opticks, seems to have run through all the Conick Sections in shaping and figuring its Parts; and Dioptrical Artists have almost ground both their Brain and Tools in pieces to find out the Arches* and Convexities of its prime parts, and are yet at a loss to find their true Figurations, whereby to advance the Fabrick of their *Telescopes* and *Microscopes*: which practical part of Opticks is but yet in the rise. But if it run on as successfully as it has begun, our Posterity may come by Glasses to out-see the Sun, and Discover Bodies in the remote Universe, that lie in Vortexes,* beyond the reach of the great Luminary. At present let us be content with what our *Microscope* demonstrates; and the former Observations, I am sure, will give all ingenious persons great occasion both to admire Nature's Anomaly in the Fabrick, as well as in the number of Eyes, which she has given to several Animals. We see the *Tunica Cornea** in most Insects is full of perforations, as if it were a *Tunica Uvea** pinked* full of Holes, and whereas perfect Animals have but one Aperture, these Insects have a thousand Pupils, and so see a Hemisphere at once. And indeed 'tis worth our consideration to think, that since their Eye is perfectly fixed, and can move no wayes; it was requisite to lattice* that Window, and supply the defect of its Motion with the multiplicity of its Apertures, that so they might see at once what we can but do at several times, our Eyes having the liberty and advantage to move every way (like Balls in Sockets) which theirs have not.

The Conclusion. To the generous VIRTUOSI, and Lovers of Experimental Philosophy*

Certainly this World was made not onely to be Inhabited but Studied and Contemplated by Man; and How few are there in the World that perform this homage due to their Creator? Who, though he hath disclaimed all Brutal, yet still accepts of a Rational Sacrifice; 'tis a Tribute we ought to pay him for being men, for it is Reason that transpeciates* our Natures, and makes us little lower than the Angels. Without the right management of this Faculty we do not so much in our kind as Beasts do in theirs, who justly obey the prescript of their Natures, and live up to the height of that instinct that Providence hath given them. But alas, How many Souls are there that never come to act beyond that of the grazing-Monarch's? *Humanum paucis vivit genus.*[9] There is a world of People indeed, and but a few Men in it; mankind is but preserv'd in a few Individuals; the greatest part of Humanity is lost in Earth, and their Souls so fixed in that grosser moity* of themselves (their Bodies) that nothing can volatilize* them, and set their Reasons at Liberty....

But it is not this numerous piece of Monstrosity (the Multitude onely) that are enemies to themselves and Learning; there is a company of men amongst the Philosophers themselves, a sort of Notional heads, whose ignorance (though varnish'd over with a little squabling Sophistry) is as great and invincible as the former. These are they that daily stuff our Libraries with their Philosophical Romances, and glut the Press with their Canting Loquacities.[10] For, instead of solid and Experimental Philosophy, it has been held accomplishment enough to graduate a Student, if he could but stiffly wrangle out a vexatious dispute of some odd Peripatetick qualities*, or the like; which (if translated into *English*) signified no more than a Heat* 'twixt two Oyster-wives in *Billingsgate*. Nay, these crimes have not onely stain'd the Common, but there are spots also to be seen even in the Purple Gowns of Learning. For it hath been a great fault, and indeed a solemn piece of Folly, even amongst the Professors and nobler sort of Philosophers, That when they have arrived to a competent height in any Art or Science, if any difficulty do arise that their Art cannot presently reach unto they instantly pronounce it a thing impossible to be done; which inconsiderable and rash censure and forestallment of their endevours does not onely stifle their own further Enquiries, but also hangs to all succeeding ages as a Scarecrow to affright them for ever approching that difficulty. Hence it is that most Arts and Sciences are branded at this day with some such ignominious Impossibility.

Thus came they to upbraid[11] Chymistry with the Alkahest, and Philosophers-Stone; Geography, with Longitudes; Geometry, with the Quadrature of a Circle; Stereometry, with the Duplication of the Cube; Trigonometry, with the Trisection of an Angle; Algebra, with the Æquation of three discontinued Numbers; Mechanicks, with a Perpetual Motion; and our own Profession with the incurability of Cancers and Quartans.* Nay, the Spring and Nepetides* in Natural Philosophy, the Doctrine of Comets in Astronomy, the Terra Incognita in Geography, the Heart's Motion in Anatomy, the Forming of Conick Sections in Dioptricks, the Various Variation in Magnetical Philosophy, are accounted as insuperable difficulties as the former, whose Causes (they say) defie all Humane Industry ever to discover them.

But besides this Intestine war and civil dissention that is 'twixt men of the same denomination and principles, there is one more general Impediment, which is an Authentick discouragement to the promotion of the Arts and Sciences, and that is The Universal Exclamation of the World's decay and approximation to its period; That both the great and little World have long since pass'd the Meridian, and, That the Faculties of the one doe fade and decay, as well as the Fabricks and Materials of the other . . . .

> [Power accepts the computation of the age of the
> world, derived from scripture, as 5,000 years, and
> argues that the "slow motion of the Sun's *Apogaeum*"
> means that the world will last another 15,000 years.]

Thus much for the *Macrocosm*. Now what decay there is in the *Microcosm*, we must be both Parties and Judges; and how far our Modern Wits have outdone the Ancient Sages, the parallel 'twixt the few Inventions of the one, and the rare Discoveries of the other, will easily determine. But the Learned *Hackwell*'s Apology[12] shall be mine at present, for not treating any further of this Subject; he having long since perform'd that Task, to the conviction of Prejudice itself.

Besides this Catholick* one, there are other Remora's* yet in the way, that have been accessory hindrances to the advancement of Learning, and that is, A diffidence and desperation of most men (nay even of those of more discerning faculties) of ever reaching to any eminent Invention; and an inveterate conceit they are possess'd with of the old Maxim, That *Nil dictum, quod non priùs dictum*:[13] by which despondency of mind, they have not onely stifled the blossoming of the Tree of Knowledge in themselves, but also have nipp'd the very Buds and Sproutings of it in others, by blazing* about the old and uncomfortable Aphorism of our *Hippocrates*,[14] of Nature's obscurity, the Life's brevity, the Senses fallacity, and the Judgement's infirmity.

Had the winged Souls of our modern Hero's been lime-twig'd with such ignoble conceptions as these, they had never flown up to those rare Inventions with which they have so enrich'd our latter dayes; we had wanted the useful Inventions of Guns, Printing, Navigation, Paper, and Sugar; we had wanted Decimal and Symbolical Arithmetick, the Analytical Algebra, the Magnetical Philosophy, the Logarithms, the Hydrargyral* Experiments, the glorious Inventions of Dioptrick Glasses, Wind-guns*, and the Noble *Boyle's*[15] Pneumatick Engine.

Nay, what strangers had we been at home, and within the circle of our own selves? We had yet never known the Mesenterical* and Thoracical *Lactea*, the Blood's Circulation, the Lymphiducts,* and other admirable Curiosities in this fabrick of our-Selves.

All which incomparable Inventions do not fully solicite, but, methinks, should inflame our endevours to attempt even Impossibilities, and to make the world know There are not difficulties enough, in Philosophy, for a vigorous and active Reason.[16] Tis a Noble resolution to begin there where all the world has ended; and an Heroick attempt to salve those difficulties (which former Philosophers accounted *impossibilities*) though but in an Ingenious Hypothesis. And, certainly, there is no Truth so abstruse, nor so far elevated out of our reach, but man's wit may raise Engines to Scale and Conquer it. Though *Democritus* his pit be never so deep[17] yet by a long *Sorites** of Observations, and chain of Deductions, we may at last fathom it, and catch hold of Truth that hath so long sitt forlorn at bottom thereof.

But these are Reaches that are beyond all those of the *Stagyrite's* Retinue,[18] the Solutions of all those former Difficulties are reserved for you (most Noble Souls, the true Lovers of Free, and Experimental Philosophy) to gratifie Posterity withall.

Henry Power (1623–1668)

You are the enlarged and Elastical Souls of the world, who, removing all former rubbish, and prejudicial resistances, do make way for the Springy Intellect to flye out into its desired Expansion. When I seriously contemplate the freedom of your Spirits, the excellency of your Principles, the vast reach of your Designs to unriddle all Nature; methinks, you have done more than men already, and may be well placed in a rank Specifically different from the rest of groveling Humanity.

And this is the Age wherein all mens Souls are in a kind of fermentation, and the spirit of Wisdom and Learning begins to mount and free it self from those drossie* and terrene Impediments wherewith it hath been so long clogg'd, and from the insipid* phlegm and *Caput Mortuum** of useless Notions, in which it has endured so violent and long a fixation.

This is the Age wherein (me-thinks) Philosophy comes in with a Spring-tide; and the Peripateticks may as well hope to stop the Current of the Tide, or (with *Xerxes*)[19] to fetter the Ocean, as hinder the overflowing of free Philosophy. Me-thinks I see how all the old Rubbish must be thrown away, and the rotten Buildings be overthrown, and carried away with so powerful an Inundation. These are the days that must lay a new Foundation of a more magnificent Philosophy, never to be overthrown, that will Empirically* and Sensibly canvass* the *Phænomena* of Nature, deducing the Causes of things from such Originals in Nature as we observe are producible by Art, and the infallible demonstration of Mechanicks: and certainly this is the way, and no other, to build a true and permanent Philosophy. For Art, being the Imitation of Nature (or Nature at Second-Hand) it is but a sensible expression of Effects, dependent on the same (though more remote Causes;) and therefore the works of the one must prove the most reasonable discoveries of the other. And to speak yet more close to the point, I think it is no Rhetorication to say That all things are Artificial, for Nature it self is nothing else but the Art of God. Then, certainly, to find the various turnings and mysterious process of this divine Art, in the management of this great Machine of the World, must needs be the proper Office of onely the Experimental and Mechanical Philosopher. For the old Dogmatists and Notional Speculators, that onely gaz'd at the visible effects and last Resultances of things, understood no more of Nature, than a rude Countrey-fellow does of the Internal Fabrick of a Watch, that onely sees the Index and Horary Circle,* and perchance hears the Clock and Alarum strike in it. But he that will give a satisfactory Account of those *Phænomena*, must be an Artificer indeed, and one well skill'd in the Wheelwork and Internal Contrivance of such Anatomical Engines.

Robert Hooke

(1635–1702)

6. *Micrographia* (1665)

Robert Hooke was educated at Westminster School and Christ Church, Oxford, where he came into contact with John Wilkins's scientific group at Wadham. Hooke acted as scientific assistant, first to Thomas Willis, then to Boyle, for whom he built an improved air-pump, his reputation as a practical scientist being so great that in 1662 he was appointed curator of experiments to the Royal Society. His job was to furnish each meeting "with three or four considerable Experiments", and for many years Hooke organized, demonstrated and lectured on the experiments performed each week. In 1664 he received a life appointment to a lectureship in mechanics, founded (but only reluctantly paid for) by Sir John Cutler; in 1665 he became Gresham professor of geometry; and in 1666 one of the three surveyors to the city of London after the great fire, together with Wren. He lived in Gresham College from 1664 until his death, was Secretary of the Royal Society from 1677 to 1682, and also acted as Librarian, and Keeper of its rareties. Hooke collected six of his works as *Lectiones Cutlerianae* (London, 1679), which included the discovery of Hooke's law of elasticity, *ut tensio sic vis*, "That is, The Power of any Spring is in the same proportion with the Tension thereof", and an important account of harmonic motion. His *Posthumous Works* were published by Richard Waller in 1705, and further *Philosophical Experiments and Observations* by William Derham in 1726.

Hooke was recognized as one of the mechanical geniuses of his age, with a remarkable ability to construct scientific instruments that were not only accurate but simple to make and use. Apart from the air-pump, in 1658 he invented a watch controlled by a spring instead of a pendulum (Huygens perfected the use of a spiral spring in 1674, but Hooke claimed priority), the wheel-barometer (still in use today), a compound microscope with a new technique of illumination, the iris diaphragm for telescopes, a wind-gauge, a sealed thermometer, a weather-clock, a hygrometer, apparatus for depth-sounding, a marine barometer, and a universal joint. He experimented and lectured on physics, chemistry, optics, biology, atlas-making, and geology, particular topics including the nature of light, gravity, combustion, and comets. He was not only city surveyor but an architect of distinction, his buildings including the Royal College of Physicians, Bedlam Hospital, Montague House, Ragley Hall, the Naval College at Greenwich (with Wren), and the Monument. Hooke never achieved the highest status as a scientist, since he was not a theorist but a practitioner, who had not advanced far in mathematics, and tended to work by intuitive understanding rather than sustained thought. Yet in his private correspondence with Newton in 1679–80 outlining the elements of orbital dynamics he acted as a catalyst, stimulating Newton to develop the inverse square relation in a coherent mathematical form. Newton acknowledged the value of Hooke's contribution to his own thinking, although relations between them – both were quick to suspect an injury and slow to forgive – were not cordial.

Micrographia was the first book to publicize the microscope. Although invented by Galileo in 1615, little use had been made of the instrument, no one having solved the problems of illuminating the object properly, or providing accurate illustrations. Hooke did both, and his technical achievement stimulated enquiry in biology, botany, and entomology, bringing microscopy to a wider public. Yet its full scientific potential was only revealed by Antoni van Leewenhoek (1632–1723) in many biological observations, such as his discovery of micro-organisms, communicated to the Royal Society in 1676.

Source: *Micrographia: or Some Physiological Descriptions of Minute Bodies made by Magnifying Glasses. With Observations and Inquiries thereupon.* (London, 1665)

The Preface.

It is the great prerogative of Mankind above other Creatures that we are not only able to *behold* the works of Nature, or barely to *sustein* our lives by them, but we have also the power of *considering*, *comparing*, *altering*, *assisting*, and *improving* them to various uses. And as this is the peculiar priviledge of humane Nature in general, so is it capable of being so far advanced by the helps of Art and Experience as to make some Men excel others in their Observations and Deductions almost as much as they do Beasts. By the addition of such *artificial Instruments* and *methods* there may be, in some manner, a reparation made for the mischiefs and imperfections mankind has drawn upon it self by negligence and intemperance, and a wilful and superstitious deserting the Prescripts and Rules of Nature, whereby every man, both from a deriv'd corruption innate and born with him, and from his breeding and converse with men, is very subject to slip into all sorts of errors.

The only way which now remains for us to recover some degree of those former perfections seems to be by rectifying the operations of the *Sense*, the *Memory*, and *Reason*, since upon the evidence, the *strength*, the *integrity* and the *right correspondence* of all these, all the light by which our actions are to be guided is to be renewed, and all our command over things is to be establisht.[1]

It is therefore most worthy of our consideration to recollect their

several defects, that so we may the better understand how to supply*
them, and by what assistances we may *inlarge* their power and *secure*
them in performing their particular duties.

As for the actions of our *Senses*, we cannot but observe them to be
in many particulars much outdone by those of other Creatures, and
when at best to be far short of the perfection they seem capable of.
And these infirmities of the Senses arise from a double cause, either
from the *disproportion of the Object to the Organ*, whereby an infinite
number of things can never enter into them, or else from *error in the
Perception*, that many things which come within their reach are not
received in a right manner.

The like frailties are to be found in the *Memory*: we often let many
things *slip* away from us which deserve to be retain'd; and of those
which we treasure up a great part is either *frivolous* or *false*; and if good
and substantial, either in tract of time *obliterated*, or at best so
overwhelmed and buried under more frothy notions that when there is
need of them they are in vain sought for.

The two main foundations being so deceivable, it is no wonder that
all the succeeding works which we build upon them, of arguing,
concluding, defining, judging and all the other degrees of Reason, are
lyable to the same imperfection, being at best either vain or uncertain.
So that the errors of the *understanding* are answerable to the two
other, being defective both in the quantity and goodness of its
knowledge; for the limits to which our thoughts are confind are small
in respect of the vast extent of Nature it self, some parts of it are *too
large* to be comprehended, and some *too little* to be perceived. And
from thence it must follow that not having a full sensation of the
Object we must be very lame and imperfect in our conceptions about
it, and in all the propositions which we build upon it; hence we often
take the *shadow* of things for the *substance*, small *appearances* for good
similitudes, similitudes for *definitions*; and even many of those which
we think to be the most solid definitions are rather expressions of our
own misguided apprehensions than of the true nature of the things
themselves. . . .

These being the dangers in the process of humane Reason the
remedies of them all can only proceed from the real, the *mechanical*,
the *experimental* Philosophy, which has this advantage over the
Philosophy of *discourse* and *disputation*, that whereas that chiefly aims
at the subtilty of its Deductions and Conclusions without much
regard to the first ground-work, which ought to be well laid on the
Sense and Memory; so this intends the right ordering of them all, and
the making them serviceable to each other.

Robert Hooke (1635–1702)

The first thing to be undertaken in this weighty work is a *watchfulness over the failings* and an *inlargement of the dominion* of the Senses.

To which end it is requisite, first, That there should be a *scrupulous* choice and a *strict examination* of the reality, constancy, and certainty of the Particulars that we admit. This is the first rise* whereon truth is to begin, and here the most severe and most impartial diligence must be imployed; the storing up of all without any regard to evidence or use will only tend to darkness and confusion. We must not therefore esteem the riches of our Philosophical treasure by the *number* only but chiefly by the weight; the most *vulgar** Instances are not to be neglected, but above all the most *instructive* are to be entertain'd; the footsteps of Nature are to be trac'd not only in her *ordinary course* but when she seems to be put to her shifts to make many *doublings* and *turnings*, and to use some kind of art in indeavouring to avoid our discovery.

The next care to be taken in respect of the Senses is a supplying of their infirmities with *Instruments*, and, as it were, the adding of *artificial Organs* to the *natural*; this in one of them has been of late years accomplisht with prodigious benefit to all sorts of useful knowledge by the invention of Optical Glasses. By the means of *Telescopes* there is nothing so *far distant* but may be represented to our view, and by the help of *Microscopes* there is nothing so *small* as to escape our inquiry, hence there is a new visible World discovered to the understanding. By this means the Heavens are open'd and a vast number of new Stars, and new Motions, and new Productions appear in them, to which all the antient Astronomers were utterly Strangers. By this the Earth it self which lyes so neer us, under our feet, shews quite a new thing to us, and in every *little particle* of its matter we now behold almost as great a variety of Creatures as we were able before to reckon up in the whole *Universe* it self.

It seems not improbable but that by these helps the subtilty of the composition of Bodies, the structure of their parts, the various texture of their matter, the instruments and manner of their inward motions, and all the other possible appearances of things may come to be more fully discovered; all which the antient *Peripateticks* were content to comprehend in two general and (unless further explain'd) useless words of *Matter* and *Form*.[2] From whence there may arise many admirable advantages towards the increase of the *Operative* and the *Mechanick* Knowledge, to which this Age seems so much inclined, because we may perhaps be inabled to discern all the secret workings of Nature, almost in the same manner as we do those that are the productions of Art, and are manag'd by Wheels, and Engines, and Springs, that were devised by humane Wit....

The truth is, the Science of Nature has been already too long made only a work of the *Brain* and the *Fancy*: It is now high time that it should return to the plainness and soundness of *Observations* on *material* and *obvious* things. It is said of great Empires That *the best way to preserve them from decay is to bring them back to the first Principles and Arts on which they did begin.* The same is undoubtedly true in Philosophy, that by wandring far away into *invisible Notions* has almost quite destroy'd it self, and it can never be recovered or continued but by returning into the same *sensible* paths* in which it did at first proceed.

If therefore the Reader expects from me any infallible Deductions or certainty of *Axioms*, I am to say for my self that those stronger Works of Wit and Imagination are above my weak Abilities; or if they had not been so, I would not have made use of them in this present Subject before me. Wherever he finds that I have ventur'd at any small Conjectures at the causes of the things that I have observed, I beseech him to look upon them only as *doubtful Problems* and *uncertain ghesses*, and not as unquestionable Conclusions or matters of unconfutable Science. I have produced nothing here with intent to bind his understanding to an *implicit* consent. I am so far from that, that I desire him not absolutely to rely upon these Observations of my eyes, if he finds them contradicted by the future Ocular Experiments of sober and impartial Discoverers....

The *Understanding* is to *order* all the inferiour services of the lower Faculties; but yet it is to do this only as a *lawful Master*, and not as a *Tyrant*. It must not *incroach* upon their Offices, nor take upon it self the employments which belong to either of them. It must *watch* the irregularities of the Senses but it must not go before them or *prevent** their information. It must *examine, range,* and *dispose* of the bank* which is laid up in the Memory: but it must be sure to make *distinction* between the *sober* and *well collected heap**, and the *extravagant Idea's* and *mistaken Images* which there it may sometimes light upon. So many are the *links* upon which the true Philosophy depends, of which, if any one be *loose* or *weak* the whole *chain* is in danger of being dissolv'd. It is to *begin* with the Hands and Eyes and to *proceed* on through the Memory, to be *continued* by the Reason; nor is it to stop there, but to *come about* to the Hands and Eyes again, and so by a *continual passage round* from one Faculty to another it is to be maintained in life and strength, as much as the body of man is by the *circulation* of the blood through the several parts of the body, the Arms, the Fat, the Lungs, the Heart and the Head.

If once this method were followed with diligence and attention there is nothing that lyes within the power of human Wit (or which is far more effectual) of human Industry, which we might not compass; we might not only hope for Inventions to equalize those of *Copernicus, Galileo, Gilbert, Harvy*, and of others whose Names are almost lost, that were the Inventors of *Gun-powder*, the *Seamans Compass, Printing, Etching, Graving, Microscopes &c.* but multitudes that may far exceed them: for even those discoveries seem to have been the products of some such method, though but imperfect. What may not be therefore expected from it if thoroughly prosecuted? *Talking* and *contention of Arguments* would soon be turn'd into *labours*; all the fine *dreams* of Opinions, and *universal metaphysical natures*, which the luxury of subtil Brains has devis'd, would quickly vanish, and give place to *solid Histories, Experiments* and *Works*. And as at first mankind *fell* by *tasting* of the forbidden Tree of Knowledge, so we, their Posterity, may be in part *restor'd* by the same way, not only by *beholding* and *contemplating* but by *tasting* too those fruits of Natural knowledge that were never yet forbidden. . . .

'Tis not unlikely but that there may be yet invented several other helps for the eye, as much exceeding those already found as those do the bare eye, such as by which we may perhaps be able to discover *living Creatures* in the Moon or other Planets, the *figures* of the compounding Particles of matter, and the particular *Schematisms** and *Textures* of Bodies.

And as *Glasses* have highly promoted our *seeing* so 'tis not improbable but that there may be found many *Mechanical Inventions* to improve our other Senses, of *hearing, smelling, tasting, touching.* 'Tis not impossible to hear a *whisper* a *furlongs* distance, it having been already done; and perhaps the nature of the thing would not make it more impossible though that furlong should be ten times multiply'd. And though some famous Authors have affirm'd it impossible to hear through the *thinnest plate* of *Muscovy-glass,** yet I know a way by which 'tis easie enough to hear one speak through a *wall a yard thick.* It has not been yet thoroughly examin'd how far *Otocousticons** may be improv'd, nor what other wayes there may be of *quickning* our hearing or *conveying* sound through *other bodies* than the *Air*: for that that is not the only *medium* I can assure the Reader that I have, by the help of a *distended wire*, propagated the sound to a very considerable distance in an *instant*, or with as seemingly quick a motion as that of light, at least incomparably swifter than that which at the same time was propagated through the Air; and this not only in a straight line or direct, but in one bended in many angles. . . .

What a prodigious variety of Inventions in *Anatomy* has this latter Age afforded,[3] even in our own Bodies, in the very *Heart* by which we live, and the *Brain* which is the seat of our knowledge of other things? witness all the excellent Works of *Pecquet*, *Bartholinus*, *Billius*, and many others; and at home, of Doctor *Harvy*, Doctor *Ent*, Doctor *Willis*, Doctor *Glisson*. In *Celestial Observations* we have far exceeded all the Antients, even the *Chaldeans* and *Egyptians* themselves, whose *vast Plains*, *high Towers*, and *clear Air* did not give them so great advantages over us as we have over them by our *Glasses*. By the help of which they have been very much outdone by the famous *Galileo*, *Hevelius*, *Zulichem*; and our own Countrymen, Mr. *Rook*, Doctor *Wren*, and the great Ornament of our Church and Nation, the *Lord Bishop of Exeter*. And to say no more in *Aerial Discoveries*, there has been a wonderful progress made by the *Noble Engine* of *the most Illustrious Mr. Boyle*, whom it becomes me to mention with all honour, not only as my particular Patron but as the *Patron* of *Philosophy* it self; which he every day *increases* by his *Labours* and *adorns* by his *Example*.[4]

The good success of all these *great Men* and many others, and the now seemingly great *obviousness* of most of their and divers other Inventions, which from the beginning of the world have been, as 'twere, trod on,* and yet not minded till these last *inquisitive* Ages (an Argument that there may be yet behind multitudes of the like), puts me in mind to recommend such Studies, and the prosecution of them by such methods, to the *Gentlemen* of our Nation, whose *leisure* makes them fit to *undertake*, and the *plenty* of their fortunes to *accomplish* extraordinary things in this way. And I do not only propose this kind of *Experimental Philosophy* as a matter of high *rapture* and *delight* of the mind, but even as a *material* and *sensible Pleasure*. So vast is the *variety of Objects* which will come under their Inspections, so many *different wayes* there are of *handling* them, so great is the *satisfaction of finding* out *new things* that I dare compare the *contentment* which they will injoy not only to that of *contemplation*, but even to that which most men prefer of *the very Senses themselves*. . . .

What each of the delineated* Subjects are, the following descriptions anext to each will inform, of which I shall here, only once for all, add That in divers of them the Gravers* have pretty well follow'd my directions and draughts; and that in making of them I indeavoured (as far as I was able) first to discover the true appearance, and next to make a plain representation of it. This I mention the rather because of these kind of Objects there is much more difficulty to discover the true shape than of those visible to the naked eye, the same Object seeming quite differing in one position to the Light, from what it really is and may be discover'd in another. And therefore I never began to make any draught before by many examinations in several lights, and in several positions to those lights, I had discover'd the true form. For it is exceeding difficult in some Objects to distinguish between a *prominency* and a *depression*, between a *shadow* and a *black stain*, or a *reflection* and a *whiteness in the colour*. Besides, the transparency of most Objects renders them yet much more difficult then if they were *opacous*. The Eyes of a Fly in one kind of light appear almost like a Lattice, drill'd through with abundance of small holes; which probably may be the Reason why the Ingenious *Dr. Power* seems to suppose them such.[5] In the Sunshine they look like a Surface cover'd with golden Nails; in another posture like a Surface cover'd with Pyramids; in another with Cones; and in other postures of quite other shapes; but that which exhibits the best is the Light collected* on the Object, by those means I have already describ'd.[6]

And this was undertaken in prosecution of the Design which the *ROYAL SOCIETY* has propos'd to it self. For the Members of the Assembly having before their eys so many *fatal* Instances of the errors and falshoods in which the greatest part of mankind has so long wandred because they rely'd upon the strength of humane Reason alone, have begun anew to correct all *Hypotheses* by sense, as Seamen do their *dead Reckonings** by *Cælestial Observations*; and to this purpose it has been their principal indeavour to *enlarge* & *strengthen* the *Senses* by *Medicine* and by such *outward Instruments* as are proper for their particular works. By this means they find some reason to suspect that those effects of Bodies which have been commonly attributed to *Qualities*, and those confess'd to be *occult*,[7] are perform'd by the small *Machines* of Nature, which are not to be discern'd without these helps, seeming the meer products of *Motion, Figure*, and *Magnitude* . . . .

And the ends of all these Inquiries they intend to be the *Pleasure* of Contemplative minds, but above all the *ease and dispatch* of the labours of mens hands. They do indeed neglect no opportunity to bring all the *rare* things of Remote Countries within the compass of their knowledge and practice. But they still acknowledg their *most useful* Informations to arise from *common* things, and from *diversifying* their most *ordinary* operations upon them. They do not wholly reject Experiments of meer *light* and *theory*; but they principally aim at such whose Applications will *improve and facilitate* the present way of *Manual Arts*[8]. And though some men, who are perhaps taken up about less honourable Employments, are pleas'd to censure their proceedings, yet they can shew more *fruits* of their first three years wherein they have assembled than any other *Society* in *Europe* can for a much larger space of time. 'Tis true, such undertakings as theirs do commonly meet with small incouragement, because men are generally rather taken with the *plausible* and *discursive* than the *real* and the solid part of Philosophy. Yet by the good fortune of their institution, in an Age of all others the most *inquisitive*, they have been assisted by the *contribution* and *presence* of very many of the chief *Nobility* and *Gentry*, and others who are some of the *most considerable* in their several Professions. But that that yet farther convinces me of the *Real esteem* that the more *serious* part of men have of this *Society* is that several *Merchants*, men who act in earnest (whose Object is *meum & tuum*,* that great *Rudder* of humane affairs) have adventur'd considerable sums of *Money* to put in practice what some of our Members have contrived, and have continued *stedfast* in their good opinions of such Indeavours when not one of a hundred of the vulgar have believed their undertakings feasable....

[Hooke praises Sir John Cutler for endowing a Royal Society Lectureship "for the promotion of Mechanick Arts".]

But to return to my Subject from a digression which, I hope, my Reader will pardon me, seeing the Example is so rare that I can make no more such digressions. If these my first Labours shall be any wayes useful to inquiring men I must attribute the incouragement and promotion of them to a very *Reverend* and *Learned Person*, of whom this ought in justice to be said, *That there is scarce any one Invention which this Nation has produc'd in our Age but it has some way or other been set forward by his assistance.* My Reader, I believe, will quickly ghess that it is *Dr. Wilkins* that I mean. He is indeed a man born for the *good* of *mankind* and for the *honour* of his *Country*. In the *sweetness* of whose *behaviour*, in the *calmness* of his *mind*, in the *unbounded goodness* of his *heart*, we have an evident Instance what the true and the *primitive unpassionate Religion* was before it was *sowred* by particular *Factions*. In a word, his *Zeal* has been so *constant* and *effectual* in advancing all good and profitable *Arts* that, as one of the Antient *Romans* said of *Scipio, That he thanked God that he was a* Roman, *because whereever* Scipio *had been born there had been the seat of the Empire of the world,* So may I thank God that *Dr. Wilkins* was an *Englishman*, for whereever he had lived there had been the chief Seat of *generous Knowledge* and *true Philosophy*. To the truth of this there are so many worthy men living that will subscribe, that I am confident what I have here said will not be look'd upon by any ingenious Reader as a *Panegyrick*, but only as a *real testimony*.

By the Advice of this *Excellent man* I first set upon this Enterprise, yet still came to it with much *Reluctancy* because I was to follow the footsteps of so eminent a Person as *Dr. Wren*, who was the first that attempted any thing of this nature; whose original draughts do now make one of the Ornaments of that great Collection of Rarities in the *Kings Closet*.[9] This *Honor* which his first beginnings of this kind have receiv'd, to be admitted into the most famous place of the world, did not so much *incourage* as the *hazard* of coming after *Dr. Wren* did *affright* me; for of him I must affirm, that since the time of *Archimedes* there scarce ever met in one man, in so great a perfection, such a *Mechanical Hand* and so *Philosophical* a *Mind*.

But at last, being assured both by *Dr. Wilkins* and *Dr. Wren* himself that he had given over his intentions of prosecuting it, and not finding that there was any else design'd the pursuing of it, I set upon this undertaking, and was not a little incourag'd to proceed in it by the Honour the *Royal Society* was pleas'd to favour me with in approving of those draughts (which from time to time as I had an opportunity of describing) I presented to them. And particularly by the Incitements of divers of those Noble and excellent Persons of it which were my more especial Friends, who were not less urgent with me for the publishing than for the prosecution of them.

After I had almost compleated these Pictures and Observations (having had divers of them ingraven and was ready to send them to the Press) I was inform'd that the Ingenious Physitian *Dr. Henry Power* had made several *Microscopical* Observations, which had I not afterwards, upon our interchangably viewing each others Papers, found that they were for the most part differing from mine, either in the Subject it self or in the particulars taken notice of; and that his design was only to print Observations without Pictures, I had even then *suppressed* what I had so far proceeded in. But being further *excited** by several of my Friends, in complyance with their opinions, that it would not be unacceptable to several inquisitive Men, and hoping also that I should thereby discover something New to the World, I have at length cast in my Mite into the vast Treasury of *A Philosophical History*. And it is my *hope*, as well as *belief*, that these my *Labours* will be no more comparable to the *Productions* of many other *Natural Philosophers* who are now every where busie about *greater*

Plate 2 Hooke's microscope, with the attachment he perfected for illuminating the
object; from *Micrographia* (1667)

things, than my *little Objects* are to be compar'd to the greater and
more beautiful *Works of Nature*, A Flea, a Mite, a Gnat, to an Horse,
an Elephant, or a Lyon.

OBSERV. I. OF THE POINT OF A SHARP SMALL NEEDLE.

As in *Geometry*, the most natural way of beginning is from a
Mathematical *point*, so is the same method in Observations and
Natural history the most genuine, simple, and instructive. We must
first endevour to make *letters* and draw *single* strokes true before we
venture to write whole *Sentences*, or to draw large *Pictures*. And in
Physical Enquiries we must endevour to follow Nature in the more
plain and *easie* ways she treads in the most *simple* and *uncompounded*
bodies, to trace her steps and be acquainted with her manner of
walking there, before we venture our selves into the multitude of

meanders she has in *bodies of a more complicated* nature; lest, being unable to distinguish and judge of our way we quickly lose both *Nature* our Guide and *our selves* too, and are left to wander in the *labyrinth* of groundless opinions; wanting both *judgment*, that *light*, and *experience*, that *clew* which should direct our proceedings.

We will begin these our Inquiries therefore with the Observations of Bodies of the most *simple nature* first, and so gradually proceed to those of a more *compounded* one. In prosecution of which method we shall begin with a *Physical point*; of which kind the *Point of a Needle* is commonly reckon'd for one; and is indeed for the most part made so sharp that the naked eye cannot distinguish any parts of it. It very easily pierces, and makes its way through all kind of bodies softer than it self. But if view'd with a very good *Microscope*, we may find that the *top* of a Needle (though as to the sense very *sharp*) appears a *broad*, *blunt*, and very *irregular* end; not resembling a Cone, as is imagin'd, but onely a piece of a tapering body, with a great part of the top remov'd or deficient. The Points of Pins are yet more blunt, and the Points of the most curious Mathematical Instruments do very seldome arrive at so great a sharpness; how much therefore can be built upon demonstrations made only by the productions of the Ruler and Compasses he will be better able to consider that shall but view those *points* and *lines* with a *Microscope*.

Now though this point be commonly accounted the sharpest (whence when we would express the sharpness of a point the most *superlatively* we say, As sharp as a Needle) yet the *Microscope* can afford us hundreds of Instances of Points many thousand times sharper: such as those of the *hairs* and *bristles* and *claws* of multitudes of *Insects*; the *thorns*, or *crooks*,* or *hairs of leaves* and other small vegetables; nay, the ends of the *stiriæ** or small *parallelipipeds** of *Amianthus** and *alumen plumosum*;* of many of which, though the Points are so sharp as not to be visible, though view'd with a *Microscope* (which magnifies the Object, in bulk, above a million of times) yet I doubt not but were we able *practically* to make *Microscopes* according to the *theory* of them, we might find hills, and dales, and pores, and a sufficient bredth or expansion to give all those parts elbow-room, even in the blunt top of the very Point of any of these so very sharp bodies. For certainly the *quantity* or extension* of any body may be *Divisible in infinitum*, though perhaps not the *matter*.

But to proceed: The Image we have here exhibited in the first Figure was the top of a small and very sharp Needle, whose point *a a* nevertheless appear'd through the *Microscope* above a quarter of an inch broad, not round nor flat but *irregular* and *uneven*; so that it

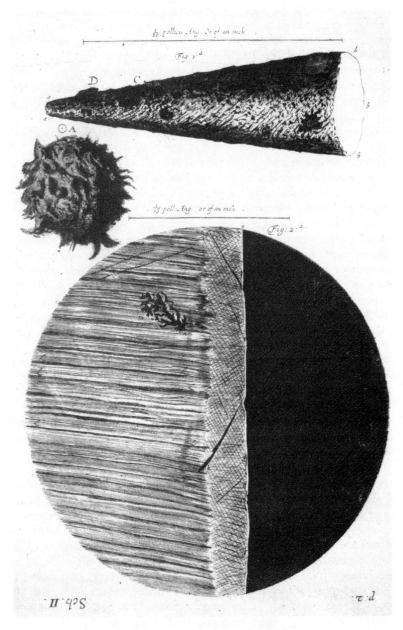

Plate 3   The point of a needle (Hooke, Scheme II)

III

seem'd to have been big enough to have afforded a hundred armed Mites room enough to be rang'd by each other without endangering the breaking one anothers necks by being thrust off on either side. The surface of which, though appearing to the naked eye very smooth, could not nevertheless hide a multitude of holes and scratches and ruggednesses from being discover'd by the *Microscope* to invest it, several of which inequalities (as A, B, C seem'd *holes* made by some small specks of *Rust*; and D some *adventitious* body* that stuck very close to it) were *casual*. All the rest that roughen the surface were onely so many marks of the rudeness and bungling of *Art*. So unaccurate is it in all its productions, even in those which seem most neat, that if examin'd with an organ more acute than that by which they were made, the more we see of their *shape* the less appearance will there be of their *beauty*: whereas in the works of *Nature* the deepest Discoveries shew us the greatest Excellencies. An evident Argument that he that was the Author of all these things was no other than *Omnipotent*; being able to include as great a variety of parts and contrivances in the yet smallest Discernable Point as in those vaster bodies (which comparatively are called also Points) such as the *Earth*, *Sun*, or *Planets*. Nor need it seem strange that the Earth it self may be by an *Analogie* call'd a Physical Point. For as its body, though now so near us as to fill our eys and fancies with a sense of the vastness of it, may by a little Distance, and some convenient *Diminishing* Glasses, be made vanish into a scarce visible Speck or Point (as I have often try'd on the *Moon*, and (when not too bright) on the *Sun* it self). So, could a Mechanical contrivance succesfully answer our *Theory* we might see the least spot as big as the Earth itself; and Discover, as *Des Cartes* also conjectures, as great a variety of bodies in the *Moon* or *Planets* as in the *Earth*.

[FROM] OBSERV. VI

In the mean time I would not willingly be guilty of that *Error* which the thrice Noble and Learned *Verulam* justly takes notice of as such, and calls *Philosophiæ Genus Empiricum, quod in paucorum Experimentorum Angustiis & Obscuritate fundatum est.*[10] For I neither conclude from one single Experiment, nor are the Experiments I make use of all made upon one Subject: Nor wrest I any Experiment to make it *quadrare** with any preconceiv'd Notion. But on the contrary, I endeavour to be conversant in divers kinds of Experiments, and all and every one of those Trials I make the Standards or Touchstones by which I try all my former Notions, whether they hold out in weight, and measure, and touch, &c. For as that Body is no other than a

Counterfeit Gold which wants any one of the Proprieties of Gold (such as are the Malleableness, Weight, Colour, Fixtness in the Fire, Indissolubleness in *Aqua fortis*,* and the like), though it has all the other; so will all those Notions be found to be false and deceitful that will not undergo all the Trials and Tests made of them by Experiments. And therefore such as will not come up to the desired *Apex* of Perfection I rather wholly reject and take new than by piecing and patching endeavour to retain the old, as knowing such things at best to be but lame and imperfect. And this course I learned from Nature; whom we find neglectful of the old Body, and suffering its Decaies and Infirmities to remain without repair, and altogether sollicitous and careful of perpetuating the *Species* by new *Individuals*. And it is certainly the most likely way to erect a glorious Structure and Temple to *Nature*, such as she will be found (by any *zealous Votary*) to reside in, to begin to build a new upon a sure Foundation of Experiments.

OBSERV. XV. OF *KETTERING-STONE*, AND OF THE PORES OF *INANIMATE* BODIES.

This Stone which is brought from *Kettering* in *Northampton-shire* and digg'd out of a Quarry, as I am inform'd, has a grain altogether admirable,* nor have I ever seen or heard of any other stone that has the like. It is made up of an innumerable company of small bodies, not all of the same cize or shape, but for the most part not much differing from a Globular form, nor exceed they one another in Diameter above three or four times; they appear to the eye like the Cobb* or Ovary of a *Herring*, or some smaller fishes, but for the most part the particles seem somewhat less, and not so uniform. But their variation from a perfect globular ball seems to be only by the pressure of the *contiguous* bals which have a little deprest and protruded those toucht sides inward, and forc'd the other sides as much outwards beyond the limits of a Globe; just as it would happen if a heap of exactly round Balls of soft Clay were heap'd upon one another, or, as I have often seen a heap of small Globules of *Quicksilver** reduc'd to that form by rubbing it much in a glaz'd* Vessel with some slimy or sluggish liquor, such as Spittle,* when though the top of the upper Globules be very neer spherical, yet those that are prest upon by others exactly imitate the forms of these lately mention'd grains. . . .

The object through the *Microscope* appears like a *Congeries** or heap of Pibbles, such as I have often seen cast up on the shore by the working of the Sea after a great storm, or like (in shape, though not colour) a company of small Globules of Quicksilver look'd on with a *Microscope*, when reduc'd into that form by the way lately mentioned.

And perhaps this last may give some hint at the manner of the formation of the former. For supposing some *Lapidescent** substance to be generated, or some way brought (either by some commixture of bodies in the Sea it self or protruded in, perhaps, out of some *subterraneous* caverns) to the bottom of the Sea, and there remaining in the form of a liquor like Quicksilver, *heterogeneous* to the ambient *Saline* fluid, it may be by the working and tumbling of the Sea to and fro be jumbled and comminuted* into such Globules as may afterwards be hardned into Flints, the lying of which one upon another when in the Sea, being not very hard, by reason of the weight of the incompassing fluid may cause the undermost to be a little, though not much, varied from a globular Figure. But this only by the by.

After what manner this *Kettering-stone* should be generated I cannot learn, having never been there to view the place and observe the circumstances; but it seems to me from the structure of it to be generated from some substance once more fluid, and afterwards by degrees growing harder, almost after the same manner as I supposed the generation of Flints to be made.

But whatever were the cause of its curious texture we may learn this information from it, that even in those things which we account vile, rude, and coorse, Nature has not been wanting to shew abundance of curiosity* and excellent Mechanisme.*

We may here find a Stone, by help of a *Microscope*, to be made up of abundance of small Balls which do but just touch each other, and yet there being so many contacts they make a firm hard mass, or a Stone much harder than Free-stone.*

Next, though we can by a *Microscope* discern so curious a shape in the particles, yet to the naked eye there scarce appears any such thing; which may afford us a good argument to think that even in those bodies also whose *texture* we are not able to discern, though help'd with *Microscopes*, there may be yet *latent* so curious a *Schematisme* that it may abundantly satisfie the curious searcher who shall be so happy as to find some way to discover it.

Next, we here find a Stone, though to the naked eye a very close one, yet every way perforated with innumerable pores, which are nothing else but the *interstitia** between those multitudes of minute globular particles that compose the bulk it self; and these pores are not only discover'd by the *Microscope*, but by this contrivance. . . .

I must not here omit to take notice that in this body there is not a *vegetative** faculty that should so contrive this structure for any peculiar use of *Vegetation* or growth, whereas in the other instances of vegetable porous bodies there is an *anima*, or *forma informans*,[11] that

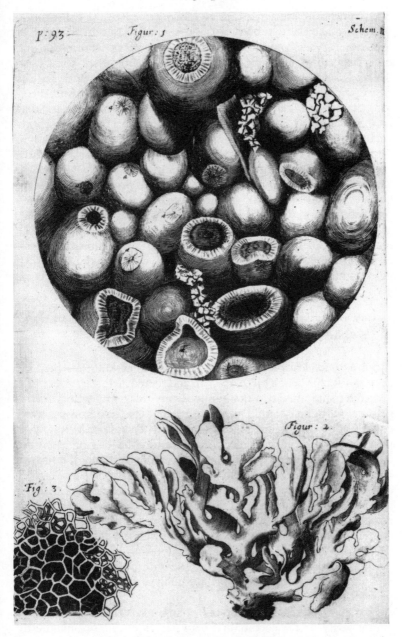

Plate 4   Kettering stone (Hooke, Scheme IX)

does contrive all the Structures and *Mechanismes* of the constituting body to make them subservient and usefull to the great Work or Function they are to perform. And so I ghess the pores in Wood and other vegetables, in bones and other Animal substances, to be as so many channels provided by the Great and Alwise Creator for the conveyance of appropriated* juyces to particular parts. And therefore, that this may tend or be pervious* all towards one part, and may have impediments, as valves or the like to any other; but in this body we have very little reason to suspect there should be any such design, for it is equally pervious every way, not onely forward but backwards, and side-ways, and seems indeed much rather to be *Homogeneous* or similar to those pores which we may with great probability believe to be the channels of *pellucid** bodies, not directed, or more open any one way than any other, being equally pervious every way. And, according as these pores are more or greater in respect of the *interstitial* bodies, the more transparent are the so constituted concretes; and the smaller those pores are, the weaker is the *Impulse* of light communicated through them, though the more quick be the progress...

OBSERV. XVIII. OF THE *SCHEMATISME** OR *TEXTURE* OF *CORK*, AND OF THE CELLS AND PORES OF SOME OTHER SUCH FROTHY BODIES.

I took a good clear piece of Cork, and with a Pen-knife sharpen'd as keen as a Razor I cut a piece of it off, and thereby left the surface of it exceeding smooth. Then examining it very dilegently with a *Microscope* me thought I could perceive it to appear a little porous; but I could not so plainly distinguish them as to be sure that they were pores, much less what Figure they were of. But judging from the lightness and yielding quality of the Cork that certainly the texture could not be so curious* but that possibly, if I could use some further diligence, I might find it to be discernable with a *Microscope*, I with the same sharp Penknife cut off from the former smooth surface an exceeding thin piece of it, and placing it on a black object Plate, because it was it self a white body, and casting the light on it with a deep *plano-convex Glass*, I could exceeding plainly perceive it to be all perforated and porous, much like a Honey-comb, but that the pores of it were not regular; yet it was not unlike a Honey-comb in these particulars.

First, in that it had a very little solid substance in comparison of the empty cavity that was contain'd between, as does more manifestly appear by the Figure A and B of the X I. *Scheme*, for the *Interstitia* or

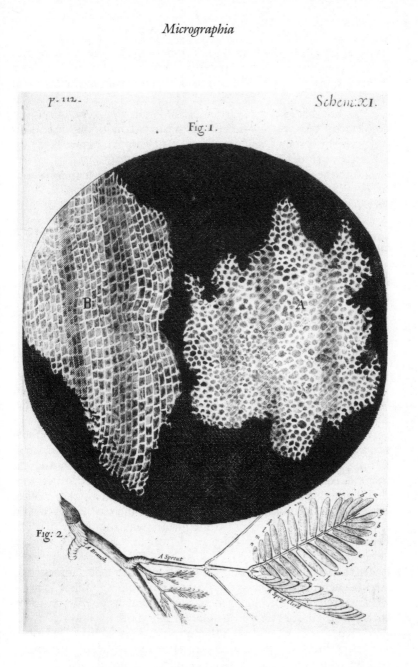

Fig: 1.

B

A

Fig: 2.

A Branch

A Sprout

A Sprig Clerk

Plate 5   Cork (Hooke, Scheme XI)

117

walls (as I may so call them) or partitions of those pores were neer as thin in proportion to their pores as those thin films of Wax in a Honey-comb (which enclose and constitute the *sexangular cells*) are to theirs.

Next, in that these pores or cells[12] were not very deep, but consisted of a great many little Boxes separated out of one continued long pore by certain *Diaphragms*, as is visible by the Figure B, which represents a sight of those pores split the long-ways.

I no sooner discern'd these (which were indeed the first *microscopical* pores I ever saw, and perhaps that were ever seen, for I had not met with any Writer or Person that had made any mention of them before this) but me thought I had with the discovery of them presently hinted to me the true and intelligible reason of all the *Phænomena* of Cork; As,

First, if I enquir'd why it was so exceeding light a body? my *Microscope* could presently inform me that here was the same reason evident that there is found for the lightness of froth, an empty Honey-comb, Wool, a Spunge, a Pumice-stone, or the like; namely, a very small quantity of a solid body extended into exceeding large dimensions.

Next, it seem'd nothing more difficult to give an intelligible reason why Cork is a body so very unapt to suck and drink in Water, and consequently preserves it self floating on the top of Water, though left on it never so long: and why it is able to stop and hold air in a Bottle, though it be there very much condens'd and consequently presses very strongly to get a passage out, without suffering the least bubble to pass through its substance. For as to the first, since our *Microscope* informs us that the substance of Cork is altogether fill'd with Air, and that that Air is perfectly enclosed in little Boxes or Cells distinct from one another, It seems very plain why neither the Water nor any other Air can easily insinuate it self into them, since there is already within them an *intus existens*,[13] and consequently why the pieces of Cork become so good floats for Nets, and stopples* for Vials, or other close Vessels.

And thirdly, if we enquire why Cork has such a springiness and swelling nature when compress'd? and how it comes to suffer so great a compression or seeming penetration of dimensions, so as to be made a substance as heavie again and more, bulk for bulk, as it was before compression, and yet suffer'd to return is found to extend it self again into the same space? Our *Microscope* will easily inform us that the whole mass consists of an infinite company of small Boxes or Bladders of Air, which is a substance of a springy nature, and that will suffer a considerable condensation* (as I have several times found by divers trials, by which I most evidently condens'd it into less than a twentieth part of its usual dimensions neer the Earth, and that with no other strength than that of my hands, without any kind of forcing Engine, such as Racks, Leavers, Wheels, Pullies or the like – but this onely by and by) and besides, it seems very probable that those very films or sides of the pores have in them a springing quality as almost all other kind of Vegetable substances have, so as to help to restore themselves to their former position.

And could we so easily and certainly discover the *Schematisme* and *Texture* even of these films, and of several other bodies as we can these of Cork; there seems no probable reason to the contrary but that we might as readily render the true reason of all their *Phænomena*; as namely, what were the cause of the springiness and toughness of some, both as to their flexibility and restitution, What of the friability* or brittleness of some others, and the like. But till such time as our *Microscope* or some other means enable us to discover the true *Schematism* and *Texture* of all kinds of bodies, we must grope, as it were, in the dark, and onely ghess at the true reasons of things by similitudes and comparisons.

But to return to our Observation. I told* several lines of these pores, and found that there were usually about threescore of these small Cells placed end-ways in the eighteenth part of an Inch in length, whence I concluded there must be neer eleven hundred of them, or somewhat more than a thousand in the length of an Inch, and therefore in a square Inch above a Million, or 1,166,400, and in a Cubick Inch, above twelve hundred Millions, or 1,259,712,000, a thing almost incredible did not our *Microscope* assure us of it by ocular demonstration; nay, did it not discover to us the pores of a body, which were they *diaphragm'd* like those of Cork, would afford us in one Cubick Inch more than ten times as many little Cells, as is evident in several charr'd* Vegetables. So prodigiously curious are the works of Nature that even these conspicuous pores of bodies, which seem to be the channels or pipes through which the *Succus nutritius* or natural juices of Vegetables are convey'd, and seem to correspond to the veins, arteries and other Vessels in sensible creatures, that these pores I say, which seem to be the Vessels of nutrition to the vastest body in the World, are yet so exceeding small that the *Atoms* which *Epicurus** fancy'd would go neer to prove too bigg to enter them, much more to constitute a fluid body in them. And how infinitely smaller then must be the Vessels of a Mite, or the pores of one of the those little Vegetables I have discovered to grow on the back-side of a Rose-leaf, and shall anon more fully describe, whose bulk is many millions of times less than the bulk of the small shrub it grows on; and even that shrub many millions of times less in bulk than several trees (that have heretofore grown in *England*, and are this day flourishing in other hotter Climates, as we are very credibly inform'd) if at least the pores of this small Vegetable should keep any such proportion to the body of it as we have found these pores of other Vegetables to do to their bulk. But of these pores I have said more elsewhere.

To proceed then, Cork seems to be by the transverse constitution of the pores a kind of *Fungus* or Mushrome, for the pores lie like so many Rays tending from the center or pith of the tree outwards; so that if you cut off a piece from a board of Cork transversly, to the flat* of it, you will, as it were, split the pores, and they will appear just as they are express'd in the Figure B of the XI. *Scheme*. But if you shave off a very thin piece from this board parallel to the plain of it you will cut all the pores transversly, and they will appear almost as they are express'd in the Figure A, save onely the solid *Interstitia* will not appear so thick as they are there represented.

So that Cork seems to suck its nourishment from the subjacent* bark of the Tree immediately, and to be a kind of excrescence or a substance distinct from the substances of the entire Tree, something *analogous* to the Mushrome or Moss on other Trees, or to the hairs on Animals. And having enquir'd into the History of Cork I find it reckoned as an excrescency of the bark of a certain Tree, which is distinct from the two barks that lie within it, which are common also to other trees; That 'tis some time before the Cork that covers the young and tender sprouts comes to be discernable; That it cracks, flaws, and cleaves into many great chaps,* the bark underneath remaining entire: That it may be separated and remov'd from the Tree, and yet the two under-barks (such as are also common to that with other Trees) not at all injur'd but rather helped and freed from an external injury....

Nor is this kind of Texture peculiar to Cork only; for upon examination with my *Microscope* I have found that the pith of an Elder, or almost any other Tree, the inner pulp or pith of the Cany hollow stalks of several other Vegetables: as of Fennel, Carrets, Daucus,* Bur-docks,* Teasels,* Fearn, some kinds of Reeds, &c. have much such a kind of *Schematisme* as I have lately shewn that of Cork, save onely that here the pores are rang'd the long-ways, or the same ways with the length of the Cane, whereas in Cork they are transverse.

The pith also that fills that part of the stalk of a Feather that is above the Quill has much such a kind of texture, save only that which way soever I set this light substance the pores seem'd to be cut transversly; so that I ghess this pith which fills the Feather not to consist of abundance of long pores separated with Diaphragms, as Cork does, but to be a kind of solid or hardned froth, or a *congeries** of very small bubbles consolidated in that form into a pretty stiff as well as tough concrete,* and that each Cavern, Bubble, or Cell is distinctly separate from any of the rest without any kind of hole in the encompassing films, so that I could no more blow through a piece of this kinde of substance than I could through a piece of Cork, or the sound pith of an Elder.

But though I could not with my *Microscope*, nor with my breath, nor any other way I have yet try'd discover a passage out of one of those cavities into another, yet I cannot thence conclude that therefore there are none such, by which the *Succus nutritius*, or appropriate juices of Vegetables may pass through them; for in several of those Vegetables, whil'st green, I have with my *Microscope* plainly enough discover'd these Cells or Poles fill'd with juices, and by degrees sweating them out: as I have also observed in green Wood all those long *Microscopical* pores which appear in Charcoal perfectly empty of anything but Air.

Now, though I have with great diligence endeavoured to find whether there be any such thing in those *Microscopical* pores of Wood or Piths as the *Valves* in the heart, veins, and other passages of Animals, that open and give passage to the contain'd fluid juices one way, and shut themselves and impede the passage of such liquors back again, yet have I not hitherto been able to say any thing positive in it; though me thinks it seems very probable that Nature has in these passages, as well as in those of Animal bodies, very many appropriated* Instruments and contrivances whereby to bring her designs and end to pass, which 'tis not improbable but that some diligent Observer, if help'd with better *Microscopes*, may in time detect....

OBSERV. XX. OF *BLUE MOULD*, AND OF THE FIRST
PRINCIPLES OF VEGETATION ARISING FROM
PUTREFACTION.

The Blue and White and several kinds of hairy mouldy spots which are observable upon divers kinds of *putrify'd* bodies, whether Animal substances, or Vegetable, such as the skin, raw or dress'd,* flesh, bloud, humours, milk, green Cheese, &c. or rotten sappy Wood, or Herbs, Leaves, Barks, Roots, &c. of Plants, are all of them nothing else but several kinds of small and variously figur'd* Mushroms, which, from convenient materials in those *putrifying* bodies are, by the concurrent heat of the Air, excited to a certain kind of vegetation, which will not be unworthy our more serious speculation and examination, as I shall by and by shew. But first I must premise* a short description of this *Specimen* which I have added of this Tribe in the first Figure of the XII. *Scheme*, which is nothing else but the appearance of a small white spot of hairy mould, multitudes of which I found to bespeck & whiten over the red covers of a small book which, it seems, were of Sheeps-skin, that being more apt to gather mould, even in a dry and clean room, than other leathers. These spots appear'd through a good *Microscope* to be a very pretty shap'd Vegetative body, which, from almost the same part of the Leather, shot out multitudes of small long cylindrical and transparent stalks, not exactly streight, but a little bended with the weight of a round and white knob* that grew on the top of each of them; many of these knobs I observ'd to be very round, and of a smooth surface, such as A A, &c. others smooth likewise but a little oblong, as B; several of them a little broken, or cloven with chops* at the top, as C; others flitter'd* as 'twere, or flown all to pieces, as D D. The whole substance of these pretty bodies was a very tender* constitution*, much like the substance of the softer kind of common white Mushroms, for by touching them with a Pin I found them to be brused and torn. They seem'd each of them to have a distinct root of their own, for though they grew neer together in a cluster yet I could perceive each stem to rise out of a distinct part or pore of the Leather. Some of these were small and short, as seeming to have been but newly sprung up, of these the balls were for the most part round; others were bigger and taller, as being perhaps of a longer growth, and of these, for the most part, the heads were broken, and some much wasted, as E. What these heads contain'd I could not perceive; whether they were knobs and flowers, or seed cases I am not able to say, but they seem'd most likely to be of the same nature with those that grow on Mushroms, which they did, some of them, not a little resemble.

Both their smell and taste, which are active enough to make a sensible* impression upon those organs, are unpleasant and noisome.

I could not find that they would so quickly be destroy'd by the actual flame of a Candle as at first sight of them I conceived they would be, but they remain'd intire after I had past that part of the Leather on which they stuck three or four times through the flame of a Candle; so that it seems they are not very apt to take fire, no more than the common white Mushroms are when they are sappy.

There are a multitude of other shapes of which these *Microscopical*

Plate 6   Blue mould (Hooke, Scheme XII)

Mushroms are figur'd, which would have been a long Work to have described, and would not have suited so well with my design in this Treatise; onely amongst the rest I must not forget to take notice of one that was a little like to or resembled a Spunge, consisting of a multitude of little Ramifications* almost as that body does, which indeed seems to be a kind of Water-Mushrom, of a very pretty texture, as I else-where manifest. And a second which I must not omit, because often mingled and neer adjoining to these I have describ'd, and this appear'd much like a Thicket of bushes or brambles, very much branch'd and extended, some of them, to a great length in proportion to their Diameter, like creeping brambles.

The manner of the growth and formation of this kind of Vegetable is the third head of Enquiry which, had I time, I should follow: the figure and method of Generation in this concrete seeming to me, next after the Enquiry into the formation, figuration, or chrystalization of Salts, to be the most simple, plain, and easie; and it seems to be a *medium* through which he must necessarily pass that would with any likelihood investigate the *forma informans* of Vegetables. For as I think that he shall find it a very difficult task who undertakes to discover the form of Saline crystalizations without the consideration and prescience* of the nature and reason of a Globular form, and as difficult to explicate this configuration of Mushroms without the previous consideration of the form of Salts, so will the enquiry into the forms of Vegetables be no less, if not much more difficult, without the fore-knowledge of the forms of Mushroms, these several Enquiries having no less dependance one upon another than any select number of Propositions in Mathematical Elements may be made to have.

Nor do I imagine that the skips from the one to another will be found very great, if beginning from fluidity, or body without any form, we descend gradually till we arrive at the highest form of a bruite Animal's Soul, making the steps or foundations of our Enquiry *Fluidity, Orbiculation,* *Fixation,* *Angulization* or *Crystallization, Germination* or *Ebullition,* *Vegetation, Plantanimation,* *Animation,* *Sensation, Imagination* . . . .

OBSERV. XXXVII. OF THE FEET OF *FLIES*, AND SEVERAL OTHER *INSECTS*.

The foot of a Fly (delineated in the first *Figure* of the 23. *Scheme*, which represents three joints, the two Tallons, and the two Pattens* in a flat posture; and in the second *Figure* of the same *Scheme*, which represents only one joint, the Tallons and Pattens in another posture) is a most admirable and curious contrivance, for by this the Flies are inabled to walk against the sides of Glass perpendicularly upwards, and to contain themselves in that posture as long as they please; nay, to walk and suspend themselves against the under surface of many bodies, as the ceiling of a room or the like, and this with as great a seeming facility and firmness as if they were a kind of *Antipodes* and had a *tendency* upwards, as we are sure they have the contrary, which they also evidently discover* in that they cannot make themselves so light as to stick or suspend themselves on the under surface of a Glass well polish'd and cleans'd. Their suspension therefore is wholly to be ascrib'd to some Mechanical contrivance in their feet; which, what it is, we shall in brief explain by shewing that its Mechanism consists principally in two parts, that is, first its two Claws or Tallons, and secondly, two Palms,* Pattens, or Soles.

The two Tallons are very large in proportion to the foot, and handsomly shap'd in the manner describ'd in the *Figures* by A B, and A C. The bigger part of them from A to *d d* is all hairy or brisled, but toward the top, at C and B smooth, the tops or points which seem very sharp turning downwards and inwards, are each of them mov'd on a joint at A, by which the Fly is able to open or shut them at pleasure, so that the points B and C being entered in any pores, and the Fly endeavouring to shut them, the Claws not onely draw one against another and so fasten each other, but they draw the whole foot, G G A D D forward, so that on a soft footing, the tenters* or points G G G G (whereof a Fly has about ten in each foot, to wit two in every joint) run into the pores, if they find any, or at least make their way; and this is sensible to the naked eye, in the feet of a *Chafer*,* which, if he be suffer'd to creep over the hand or any other part of the skin of ones body, does make his steps as sensible to the touch as the sight.

But this contrivance, as it often fails the *Chafer* when he walks on hard and close bodies, so would it also our Fly, though he be a much lesser and nimbler creature, and therefore Nature has furnish'd his foot with another *additament** much more curious and admirable, and that is, with a couple of Palms, Pattens or Soles D D, the structure of which is this:

From the bottom or under part of the last joint of his foot, K, arise two small thin plated horny substances, each consisting of two flat pieces, D D, which seem to be flexible, like the covers of a Book, about F F, by which means the plains* of the two sides E E do not always lie in the same plain, but may be sometimes shut closer, and so each of them may take a little hold themselves on a body; but that is

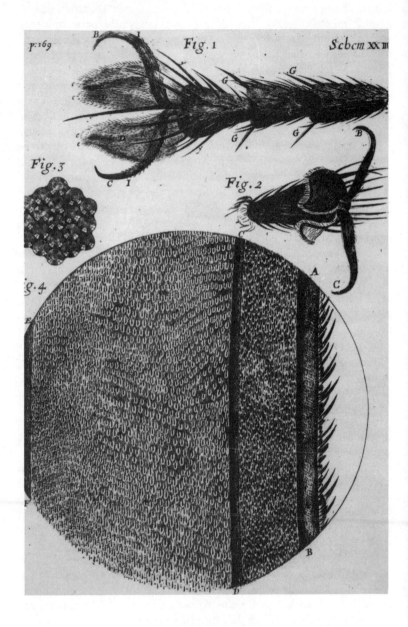

Plate 7   The foot of a fly (Hooke, Scheme XXIII)

not all, for the under sides of these Soles are all beset with small brisles, or tenters, like the Wire teeth of a Card* used for working Wool, the points of all which tend forwards. Hence the two Tallons drawing the feet forwards, as I before hinted, and these being applied to the surface of the body with all the points looking the contrary way, that is, forwards and outwards, if there be any irregularity or yielding in the surface of the body the Fly suspends it self very firmly and easily, without the access or need of any such Sponges fill'd with an imaginary *gluten** as many have, for want of good Glasses perhaps, or a troublesome and diligent examination, suppos'd.[14]

Now, that the Fly is able to walk on Glass proceeds partly from some ruggedness of the surface: and chiefly from a kind of tarnish or dirty smoaky substance which adheres to the surface of that very hard body; and though the pointed parts cannot penetrate the substance of Glass yet may they find pores enough in the tarnish, or at least make them.

This Structure I somewhat the more diligently survey'd because I could not well comprehend how, if there were such a glutinous matter in those supposed Sponges as most (that have observ'd that Object in a *Microscope*) have hitherto believ'd, how, I say, the Fly could so readily unglew and loosen its feet; and because I have not found any other creature to have a contrivance any ways like it; and chiefly, that we might not be cast upon unintelligible explications of the *Phænomena* of Nature, at least others than the true ones, where our senses were able to furnish us with an intelligible, rationall and true one.

Somewhat a like contrivance to this of Flies shall we find in most other Animals, such as all kinds of Flies and case-wing'd* creatures; nay in a Flea, an Animal abundantly smaller than this Fly. Other creatures, as Mites, the Land-Crab, &c. have onely one small very sharp Tallon at the end of each of their legs, which all drawing towards the center or middle of their body inable these exceeding light bodies to suspend and fasten themselves to almost any surface.

Which how they are able to do will not seem strange if we consider, first, how little body there is in one of these creatures compar'd to their superficies or outside, their thickness, perhaps, oftentimes not amounting to the hundredth part of an Inch. Next, the strength and agility of these creatures compar'd to their bulk being, proportionable to their bulk, perhaps an hundred times stronger than an Horse or Man. And thirdly, if we consider that Nature does always appropriate the instruments so as they are most fit and convenient to perform their offices, and the most simple and plain that possibly can be. This we may see further verify'd also in the foot of a Louse, which is very much differing from those I have been describing but more convenient and necessary for the place of its habitation, each of his leggs being footed with a couple of small claws which he can open or shut at pleasure, shap'd almost like the claws of a Lobster or Crab, but with appropriated contrivances for his peculiar use, which being to move its body to and from upon the hairs of the creature it inhabits, Nature has furnish'd one of its claws with joints, almost like the joints of a man's fingers, so as thereby it is able to encompass or grasp a hair as firmly as a man can a stick or rope.

Nor is there a less admirable and wonderfull *Mechanism* in the foot of a Spider, whereby he is able to spin, weave, and climb, or run on his curious transparent clew,* of which I shall say more in the description of that Animal.

And to conclude, we shall in all things find that Nature does not only work Mechanically but by such excellent and most compendious, as well as stupendious contrivances, that it were impossible for all the reason in the world to find out any contrivance to do the same thing that should have more convenient properties. And can any be so sottish as to think all those things the productions of chance? Certainly, either their Ratiocination must be extremely depraved or they did never attentively consider and contemplate the Works of the Al-mighty.[15]

OBSERV. XLIX. OF AN *ANT* OR *PISMIRE*.

This was a creature more troublesom to be drawn than any of the rest, for I could not, for a good while, think of a way to make it suffer its body to ly quiet in a natural posture. But whil'st it was alive, if its feet were fetter'd in Wax or Glew it would so twist and wind its body that I could not any wayes get a good view of it; and if I killed it its body was so little that I did often spoile the shape of it before I could throughly view it. For this is the nature of these minute Bodies, that as soon, almost, as ever their life is destroy'd their parts immediately shrivel and lose their beauty; and so is it also with small Plants, as I instanced before in the description of Moss. And thence also is the reason of the variations* in the beards of wild Oats and in those of Muskgrass seed, that their bodies, being exceeding small, those small variations which are made in the surfaces of all bodies almost upon every change of Air, especially if the body be porous, do here become sensible*, where the whole body is so small that it is almost nothing but surface. For as in vegetable substances, I see no great reason to think that the moisture of the Aire (that, sticking to a wreath'd* beard, does make it untwist) should evaporate, or exhale away any faster than the moisture of other bodies, but rather that the avolation* from, or access of moisture to the surfaces of bodies being much the same, those bodies become most sensible of it which have the least proportion of body to their surface. So is it also with Animal substances; the dead body of an Ant or such little creature does almost instantly shrivel and dry, and your object shall be quite another thing before you can half delineate it, which proceeds not from the extraordinary exhalation* but from the small proportion of body and juices to the usual drying of bodies in the Air, especially if warm. For which inconvenience, where I could not otherwise remove it, I thought of this expedient.

I took the creature I had design'd to delineate, and put it into a drop of very well rectified* Spirit of Wine; this I found would presently dispatch,* as it were, the Animal, and being taken out of it and lay'd on a paper, the spirit of Wine would immediately fly away and leave the Animal dry, in its natural posture, or at least in a constitution that it might easily with a pin be plac'd in what posture you desired to draw it, and the limbs would so remain without either moving or shriveling. And thus I dealt with this Ant which I have here delineated, which was one of many, of a very large kind, that inhabited under the Roots of a Tree, from whence they would sally out in great parties and make most grievous havock of the Flowers and Fruits in the ambient Garden, and return back again very expertly by the same wayes and paths they went.

It was more than half the bigness of an Earwig, of a dark brown or reddish colour, with long legs, on the hinder of which it would stand up, and raise its head as high as it could above the ground that it might stare the further about it, just after the same manner as I have also observ'd a hunting Spider[16] to do. And putting my finger towards them they have at first all run towards it till almost at it; and then they would stand round about it at a certain distance and smell, as it were, and consider whether they should any of them venture any further, till one more bold than the rest venturing to climb it, all the rest, if I would have suffered them, would have immediately followed. Many such other seemingly rational actions I have observ'd in this little Vermine with much pleasure, which would be too long to be here related; those that desire more of them may satisfie their curiosity in *Ligons* History of the *Barbadoes*.[17]

Having insnar'd several of these into a small Box, I made choice of the tallest grown among them, and separating it from the rest I gave it a Gill of Brandy, or Spirit of Wine, which after a while e'en knock'd

Plate 8   An ant (Hooke, Scheme XXXII)

him down dead drunk, so that he became moveless, though at first putting in he struggled for a pretty while very much till at last, certain bubbles issuing out of its mouth, it ceased to move. This (because I had before found them quickly to recover again if they were taken out presently) I suffered to lye above an hour in the Spirit; and after I had taken it out and put its body and legs into a natural posture, remained moveless about an hour. But then upon a sudden, as if it had been awaken out of a drunken sleep, it suddenly reviv'd and ran away. Being caught and serv'd* as before he for a while continued struggling and striving, till at last there issued several bubbles out of its mouth, and then, *tanquam animam expirasset*,[18] he remained moveless for a good while; but at length again recovering, it was again redipt, and suffered to lye some hours in the Spirit. Notwithstanding which, after it had layen dry some three or four hours, it again recovered life and motion. Which kind of Experiments, if prosecuted, which they highly deserve, seem to me of no inconsiderable use towards the invention of the *Latent Scheme* (as the Noble *Verulam* calls it) or the hidden, unknown Texture of Bodies.[19]

Of what Figure this Creature appear'd through the *Microscope* the 32. *Scheme* (though not so carefully graven as it ought) will represent to the eye, namely, That it had a large head A A, at the upper end of which were two protuberant eyes, pearl'd like those of a Fly, but smaller B B; out of the Nose or foremost part issued two horns C C, of a shape sufficiently differing from those of a blew Fly, though indeed they seem to be both the same kind of Organ, and to serve for a kind of smelling; beyond these were two indented jaws D D, which he open'd side-ways, and was able to gape them asunder very wide; and the ends of them being armed with teeth, which meeting went between each other, it was able to grasp and hold a heavy body three or four times the bulk and weight of its own body. It had only six legs, shap'd like those of a Fly, which, as I shewed before, is an Argument that it is a winged Insect, and though I could not perceive any sign of them in the middle part of its body (which seem'd to consist of three joints or pieces E F G, out of which sprung two legs) yet 'tis known that there are of them that have long wings, and fly up and down in the air.

The third and last part of its body I I I was bigger and larger than the other two, unto which it was joyn'd by a very small middle, and had a kind of loose shell, or another distinct part of its body H, which seem'd to be interpos'd and to keep the *thorax* and belly from touching.

The whole body was cas'd over with a very strong armour, and the

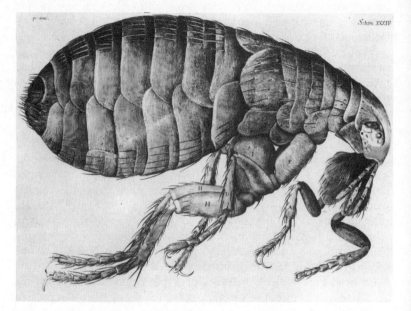

Plate 9   A flea (Hooke, Scheme XXXIV)

Belly I I I was covered likewise with multitudes of small white shining brisles; the legs, horns, head, and middle parts of its body were bestuck with hairs, also, but smaller and darker.

OBSERV. LIII. OF A *FLEA*.

The strength and beauty of this small creature, had it no other relation at all to man, would deserve a description.

For its strength, the *Microscope* is able to make no greater discoveries of it than the naked eye, but onely the curious contrivance of its leggs and joints for the exerting that strength is very plainly manifested, such as no other creature I have yet observ'd has any thing like it. For the joints of it are so adapted that he can, as 'twere, fold them short one within another and suddenly stretch or spring them out to their whole length, that is, of the fore-leggs. The part A, of the 34. *Scheme*, lies within B, and B within C, parallel to, or side by side each other; but the parts of the two next lie quite contrary, that is, D without E, and E without F, but parallel also; but the parts of the hinder leggs, G, H and I, bend one within another, like the parts of a double jointed Ruler, or like the foot, legg and thigh of a man; these six leggs he clitches* up altogether, and when he leaps springs them all out, and thereby exerts his whole strength at once. But as for the beauty of it, the *Microscope* manifests it to be all over adorn'd with a curiously polish'd suit of *sable** Armour, neatly jointed, and beset with multitudes of sharp pinns, shap'd almost like Porcupine's Quills or bright conical Steel-bodkins; the head is on either side beautify'd with a quick and round black eye K, behind each of which also appears a small cavity L, in which he seems to move to and fro a certain thin film beset with many small transparent hairs, which probably may be his ears; in the forepart of his head, between the two fore-leggs, he has two small long jointed feelers or rather smellers, M M, which have four joints, and are hairy, like those of several other creatures; between these it has a small *proboscis*, or *probe*, N N O, that seems to consist of a tube N N, and a tongue or sucker O, which I have perceiv'd him to slip in and out. Besides these it has also two chaps* or biters P P, which are somewhat like those of an Ant, but I could not perceive them tooth'd; these were shap'd very like the blades of a pair of round top'd Scizers, and were opened and shut just after the same manner. With these Instruments does this little busie Creature bite and pierce the skin and suck out the blood of an Animal, leaving the skin inflamed with a small round red spot. These parts are very difficult to be discovered because, for the most part, they lye covered between the fore-legs. There are many other particulars which, being more obvious, and affording no great matter of information, I shall pass by and refer the Reader to the Figure.

7. *On Earthquakes and Fossils* (1668)

Hooke was a pioneer in geology, the first Englishman to make any substantial contribution to this subject. Whereas fossils had previously been explained as the product of some inherent "plastic power" in nature, Hooke showed that they were the remains of natural bodies (shells, animals, wood) that had been turned into stone by a long process of coagulation or petrifaction, and had been transported from sea-level to the tops of mountains by earthquakes or floods. Hooke's thirty-year long geological observation and classification is a model of Baconian inductive method, yet he is able also to apply his new knowledge from the microscope to distinguish the polyhedral forms of crystals of different substances. His refutation of the argument that fossils are *lusus naturae*, mere sports of nature having no function or purpose, has been described as "one of the classic passages of scientific argumentation in the seventeenth century" (R. S. Westfall in *A Dictionary of Scientific Biography*, vol. VI, p. 486). The work's origin as a course of lectures accounts for both the clarity of exposition and the occasional repetitiveness.

Robert Hooke (1635–1702)

Source: *Lectures and Discourses of Earthquakes, and Subterraneous Eruptions. Explicating the Causes of the Rugged and Uneven Face of the Earth; and what Reasons may be given for the frequent finding of Shells and other Sea and Land Petrified Substances, scattered over the whole Terrestrial Superficies*, printed for the first time in *The Posthumous Works of Robert Hooke*, ed. Richard Waller (London, 1705), pp. 279–450.

A Discourse of Earthquakes.

The Introduction to the following Discourse, giving some account of its Design.

I have formerly endeavour'd to explain several Observations I had made concerning the Figure, Form, Position, Distance, Order, Motions and Operations of the Celestial Bodies, both as to themselves and one with another, and likewise with respect to the Body of the *Earth* on which we inhabit. But conceiving it may more nearly concern us to know more particularly the Constitution, Figure, Magnitude and Properties of the Body of the Earth itself, and of its several constituent Parts, I have endeavour'd to collect such Observations and Natural Histories of others as may serve to give some Light toward the making a compleat discovery of them, so far as the Power, Faculties, Organs, and other helps that Nature has furnish'd Man with, may assist us in performing and perfecting thereof.

The Subject is large, as extending as far as the whole Bulk included within the utmost limits of the Atmosphere: And 'tis not less copious and repleat with variety, as containing all the several Parts and Substances included within those Limits, namely, The aerial, watery and earthy Parts thereof, whether Superficial or Subterraneous, whether Exposed or Absconded,* whether Supraterraneal, Superterraneal, or Subterraneal, whether Elemental or Organical, Animate or Inanimate, and all the Species and Kinds of them, and all the constituent Parts of them, and the Composits constituted of them; of which also there will fall under Consideration the Artificial as well as the Natural Causes and Powers effective of things; then their Generation, Production, Augmentation, Perfection, Vertue, Power, Activity, Operation, Effect, Conservation, Duration, Declination,* Destruction, Corruption, Transformation, and in one word, the motion or progression of Nature sensibly* exprest, or any other ways discernable in each of those Species. Which Subject, if we consider as it is thus represented, doth look very like an Impossibility to be undertaken even by the whole World, to be gone through within an Age, much less to be undertaken by any particular Society or a small number of Men. The number of Natural Histories, Observations,

Experiments, Calculations, Comparisons, Deductions and Demonstrations necessary thereunto seeming to be incomprehensive and numberless: And therefore a vain Attempt, and not to be thought of till after some Ages past in making Collections of Materials for so great a Building, and the employing a vast number of Hands in making this Preparation;[20] and those of several sorts, such as Readers of History,* Criticks, Rangers* and Namesetters* of Things, Observers and Watchers of several Appearances and Progressions of Natural Operations and Perfections, Collectors of curious Productions, Experimenters and Examiners of Things by several Means and several Methods and Instruments, as by Fire, by Frost, by Menstruums,* by Mixtures, by Digestions, Putrefactions, Fermentations and Petrifactions, by Grindings, Brusings, Weighings and Measuring, Pressing and Condensing, Dilating and Expanding, Dissecting, Separating and Dividing, Sifting and Streining; by viewing with Glasses and Microscopes, Smelling, Tasting, Feeling, and various other ways of Torturing and Wracking of Natural Bodies to find out the Truth or the real Effect as it is in its Constitution or State of Being.[21]

To these may be added Registers or Compilers, such as shall Record and Express in proper Terms these Collections; add to these Examiners and Rangers of Things, such as shall distinguish and marshal them into proper Classes, and denote their Excellencies or Gradations of differing Kinds, their Perfections or Defects, what are Compleat, and what Defective and to be repeated, and the like.

So that we see the Subject of this Enquiry is very copious and large, and will afford Work enough for every Well-willer to employ his Head and Hands to contribute towards the providing Materials for so large a Fabrick and Structure as the great quantity of Materials to be collected do seem to denote. However, 'tis possible that a much less number may serve the turn, if fitly qualified* and done with Method and Design, and it may be much better and easier.

When this mighty Collection is made, what will be the use of so great a Pile*? Where will be found the Architect that shall contrive and raise the Superstructure that is to be made of them, that shall fit every one for its proper use? Till which be found they will indeed be but a heap of Confusion. Who shall find out the Experiments, the Observations, and other Remarks fit for this or that Theory? One Stone is too thick or too thin, too broad or too narrow, not of a due colour, or hardness, or grain to suit with the Design, or with some other that are duly scapled* for the purpose. This Piece of Timber is not of a right Kind, not of a sufficient Driness and Seasoning, not of a due length and bigness, but wants its Scantlings,* or is of an ill Shape for such a purpose, or was not fell'd in a due time. 'Tis Sap-rotten, or Wind-shaken, or rotten at Heart, or too frow,* and the like, for the purpose for which 'tis wanted.

The Use of predesign'd Theories, and Modules of Enquiry.

I mention this to hint only by the by that there may be use of Method in the collecting of Materials as well as in the use of them, and to shew that there may be made a Provision too great as well as too little, that there ought to be some End and Aim, some pre-design'd Module* and Theory, some Purpose in our Experiments and more particular observing of such Circumstances as are proper for that Design. And though this Honourable Society have hitherto seem'd to avoid and prohibit pre-conceived Theories and Deductions from particular and seemingly accidental Experiments, yet I humbly conceive that such, if knowingly and judiciously made, are Matters of the greatest Importance, as giving a Characteristick* of the Aim, Use, and Significancy thereof; and without which many, and possibly the most considerable Particulars, are passed over without Regard and Observation. The most part of Mankind are taken with the Prettiness or the Strangeness of the Phæ-nomena, and generally neglect the common and most obvious; whereas in truth, for the most part, they are the most considerable. And the greatest part of the Productions of Nature are to be seen every where, and by every one, though for the most part not heeded or regarded because they are so common. I could wish therefore that the Information of Experiments might be more respected than either the Novelty, the Surprizingness, the Pomp and Appearances of them.

Of figured* Stones.

The obviousness and easiness of knowing many Things in Nature has been the Cause of their being neglected, even by the more diligent and curious; which nevertheless, if well examined, do very often contain Informations of the greatest value. It has been generally noted by common as well as inquisitive* Persons that divers Stones have been found formed into the Shapes of Fishes, Shells, Fruits, Leaves, Wood, Barks, and other Vegetable and Animal Substances. We commonly know some of them exactly resembling the Shape of Things we commonly find (as the Chymists speak) in the Vegetable or Animal Kingdom; others of them indeed bearing some kind of Similitude, and agreeing in many Circumstances, but yet not exactly figured like any other thing in Nature; and yet of so curious a Shape that they easily raise both the Attention and Wonder, even of those that are less inquisitive. Of these beautifully shaped Bodies I have observed two sorts: First, some more properly natural, such as have their Figures peculiar to their Substances: Others more improperly so, that is, such as seem to receive their Shape from an external and accidental Mould.

Of Chrystals, and the like Stones, shot* into Figures.

Of the first sort are all those curiously figured Bodies of Salts, Talks,* Spars,* Crystals, Diamonds, Rubies, Amethysts, Ores, and divers other Mineral Substances wherewith the World is adorned and enriched; which I at present omit to describe as reserving them for a Second Part, they seeming to be, as it were, the elemental Figures or the *A B C* of Nature's working, the Reason of whose curious Geometrical Forms (as I may so call them) is very easily explicable Mechanically: And shall proceed to the second sort of Bodies.

Of Petrifactions.

Of these are two kinds; either first the very Substances themselves converted into Stone, such are Bones, Teeth, Shells, Fruit, Wood, Moss, Mushrooms, and divers Vegetable and Animal Substances; Or secondly, such other Mineral or Earthy Substances as Clays, Sands, Earths, Flinty Juices, &c. which have filled up, and been moulded in divers other Bodies, as Shells, Bones, Fruits, &c. but especially Shells. These, according to the Representations they bear of other Bodies, have received divers Names; . . . .

Of these I shall describe some few, because every one has not the Opportunity of seeing and examining them.

[Hooke lists, and illustrates, several types of fossil: "Snail-stones", "Snake-stones", "Nautil-Shells", "Helmet-stones", and others.]

I would to this have added the Description of a great Variety of *Echini**-shells, divers of which I have by me in the Repository of the Royal Society and others that I have met with elsewhere, but that I shall do it elsewhere. They are indeed almost infinite, but all concur in these Properties which all Helmet-stones likewise have. First, that they are distinguish'd into five Parts, by Sutures,* Ribs, and Furrows. Secondly, that they have two Vent-holes. They have divers of them also little Edges, being the Impressions of the Sutures, and divers little rows of Pins, being the Impressions* of the small Holes; and any one that will diligently and impartially examine both the Stones and

Robert Hooke (1635–1702)

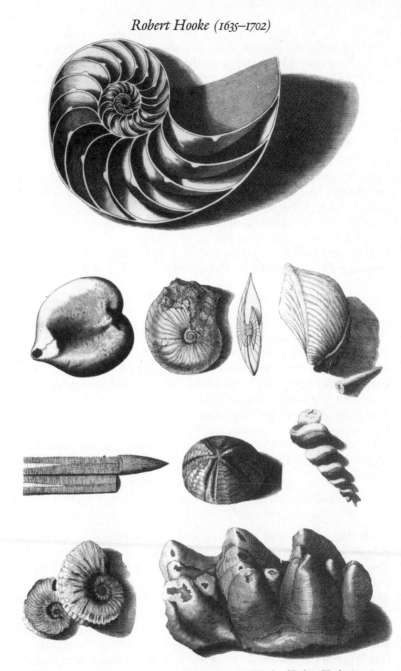

Plate 10   Fossilized stones; from *The Posthumous Works of Robert Hooke* (1705)

138

the Shells, and compare the one with the other, will, I can assure him, find greater reason to perswade him of the Truth of my Position than any I have yet urged or can well produce in Words; no Perswasions being more prevalent than those which these dumb Witnesses do insinuate....

These and the like Shapes, because many of them are curious, have so far wrought* on some Men that they have endeavoured to give us an Explication of the manner of their formation; in doing of which they have so far rambled from the true and genuine Cause of them that they have left the Matter much more difficult than they found it. Amongst the rest *Gaffarel*,[22] a *French* Writer, seems not the least mistaken, who has transferr'd them over to the Confirmation, as he thinks, of his Astrological and Magical Fancy; and thinks that as they were produced from some extraordinary Celestial Influence, and that the Aspects and Positions of the fix'd Stars and Planets conduc'd to their Generation, so that they also have in them a secret Vertue whereby they do at a distance work Miracles on things of the like shape. But these, as fantastical and groundless, I shall not spend time on at present to refute, nor on the Conjectures and Hypotheses of divers others; which though perhaps somewhat more tolerable than that I last recited, yet most of them have recourse to some vegetative or plastick Vertue* inherent in the Parts of the Earth where they were made, or in the very parcels* of which they consist, which to me seems not at all consonant to the other workings of Nature; for those more curiously carved and beautiful Forms are usually bestow'd on some vegetable or animal Body. But my Business at present shall not be so much to confute others Conjectures as to make probable some of my own; which tho' at the first hearing they may seem somewhat paradoxical, yet if the Reasons that have induced me thereunto be well consider'd and weigh'd I hope at least they may seem possible, if not more than a little probable.

Enumeration of the Phenomena.

The particular Productions of this kind that I have taken notice of my self in my own Enquiries, and which I find dispersed up and down in the Writing of others, may be reduced under some one or other of these General Heads or Propositions.

First, That there are found in most Countries of the Earth, and even in such where it is somewhat difficult to imagine (by reason of their vast distance from the Sea or Waters how they should come there) great quantities of Bodies resembling both in Substance and Shape the Shells of divers sorts of Shell-fishes; and many of them so exactly that any one that knew not whence they came would without the least scruple firmly believe them to be the Shells of such Fishes. But being found in Places so unlikely to have produced them, and not conceiving how else they should come there, they are generally believed to be real Stones form'd into these Shapes either by some plastick Vertue inherent in those Parts of the Earth, which is extravagant enough, or else by some Celestial Influence or Aspect of the Planets operating at a distance upon the yielding Matter of the Parts of the Earth, which is much more extravagant. Of this kind are all those several sorts of Oyster-shells, Cockle-shells, Muscle-shells, Periwinkle-shells, and the like, which are found in *England, France, Spain, Italy, Germany, Norway, Russia, Asia* and *Africa*, and divers other Places; of which I have very good Testimonies from Authors of good Credit.

Secondly, That there often have been, and are still daily found in other Parts of the Earth, buried below the present Surface thereof divers sorts of Bodies, besides such as I newly mention'd, resembling both in Shape, Substance, and other Properties the Parts of Vegetables, having the perfect Rind or Bark, Pith, Pores, Roots, Branches, Gums, and other constituent Parts of Wood, though in another posture, lying for the most part Horizontal, and sometimes inverted, and much differing from that of the like Vegetables when growing, and wanting also, for the most part, the Leaves, smaller Roots and Branches, the Flower and Fruit, and the like smaller Parts, which are common to Trees of that kind; of which sort is the *Lignum Fossile,** which is found in divers Parts of *England, Scotland, Ireland*, and divers Parts of *Italy, Germany*, the *Low Countries*, and indeed almost in every Country of the World.

Thirdly, That there are often found in divers other Parts of the Earth Bodies resembling the whole Bodies of Fishes, and other Animals and Vegetables, or the Parts of them, which are of a much less permanent Nature than the Shells abovemention'd, such as Fruits, Leaves, Barks, Woods, Roots, Mushrooms, Bones, Hoofs, Claws, Horns, Teeth, *&c.* but in all other Proprieties* of their Substance, save their Shape, are perfect Stones, Clays, or Earths, and seem to have nothing at all of Figure in the inward Parts of them. Of this kind are those commonly call'd Thunder-bolts, Helmet-stones, Serpentine-stones, or Snake-stones, Rams-horns, Brain-stones, Star-stones, Screw-stones, Wheel-stones, and the like.

Fourthly, That the Parts of the Earth in which these kinds have been found are some of them some hundred of Miles distant from any Sea, as in several of the Hills of *Hungary*, the Mountain *Taurus*, the *Alpes, &c.*

Fifthly, That divers of those Parts are many Scores, nay, some many Hundreds of Fathoms above the Level of the Surface of the next adjoining Sea, there having been found of them on some of the most Inland and on some of the highest Mountains in the World.

Sixthly, That divers other Parts where these Substances have been found are many Fathoms below the Level both of the Surface of the next adjoining Sea, and of the Surface of the Earth itself, they having been found buried in the bottoms of some of the deepest Mines and Wells, and inclosed in some of the hardest Rocks and toughest Metals. Of this we have continual Instances in the deapest Lead and Tin-mines, and a particular Instance in the Well dug in *Amsterdam*, where at the Depth of 99 Foot was found a Layer of Seashells mixed with Sand of 4 Foot thickness, after the Diggers had past through 7 Foot of Garden-mould, 9 Foot more of black Peat, 9 Foot more of soft Clay, 8 of Sand, 4 of Earth, 10 of Potters-clay, 4 more of Earth, 10 Foot more of Sand, upon which the Stakes or Piles of the *Amsterdam* Houses rest; then 2 Foot more of Potters-clay, and 4 of white Gravel, 5 of dry Earth, 1 of mix'd, 14 of Sand, 3 of a Sandy Clay, and 5 more of Potters-clay mix'd with Sand. Now below this Layer of Shells immediately joining to it, was a Bed of Potters-clay of no less than 102 Foot thick; but of this more hereafter.

Seventhly, That there are often found in the midst of the Bodies of very hard and close Stone, such as Marbles, Flints, *Portland*, and Purbeck-stone, *&c.* which lye upon or very near to the Surface of the Earth, great quantities of these kind of figured Bodies or Shells, and that there are many of such Stones which seem to be made of nothing else.

How the Difficulty may be solved.

These Phænomena, as they have hitherto much puzled all Natural Historians and Philosophers to give an Account of them, so in truth are they in themselves so really wonderful that 'tis not easie without making multitudes of Observations, and comparing them very diligently with the Histories and Experiments that have been already made, to fix upon a plausible Solution of them. For as on the one side it seems very difficult to imagine that Nature formed all these curious Bodies for no other End than only to play the Mimick in the Mineral Kingdom, and only to imitate what she had done for some more noble End and in a greater Perfection in the Vegetable and Animal Kingdoms; and the strictest Survey that I have made both of the Bodies themselves, and of the Circumstances obvious enough about them, do not in the least hint any thing else, they being promiscuously found of any kind of Substance, and having not the least appearance of any internal or substantial Form but only of an external or figured Superficies. As, I say, 'tis something harsh* to imagine that these thus qualified* Bodies should, by an immediate plastick* Vertue, be thus shaped by Nature contrary to her general Method of acting in all other Bodies; so on the other side it may seem at first hearing somewhat difficult to conceive how all those Bodies, if they either be the real Shells or Bodies of Fish, or other Animals or Vegetables which they represent, or an Impression left on those Substances from such Bodies, should be in such great quantities transported into Places so unlikely to have received them from any help of Man, or from any other obvious Means.

The former of these ways of solving these Phænomena I confess I cannot for the Reasons I now mention'd by any means assent unto; but the latter, tho' it has some Difficulties also, seems to me not only possible but probable.

The greatest Objections that can be made against it are, First, by what means those Shells, Woods, and other such like Substances (if they really are the Bodies they represent) should be transported to and be buried in the Places where they are found? And,

Secondly, Why many of them should be of Substances wholly differing from those of the Bodies they represent; there being some of them which represent Shells of almost all kinds of Substances, Clay, Chalk, Marble, soft Stone, harder Stone, Marble, Flint, Marchasite,* Ore, and the like.

In answer to both which, and some other of less Importance which I shall afterwards mention, give me leave to propound these following Propositions which I shall endeavour to make probable. Of these in their Order.

My first Proposition then is That all, or the greatest part of these curiously figured Bodies found up and down in divers Parts of the World, are either those Animal or Vegetable Substances they represent converted into Stone by having their Pores fill'd up with some petrifying liquid Substance, whereby their Parts are, as it were, lock'd up and cemented together in their Natural Position and Contexture; or else they are the lasting Impressions made on them at first whilst a yielding Substance by the immediate Application of such Animal or Vegetable Body as was so shaped, and that there was nothing else concurring to their Production save only the yielding of the Matter to receive the Impression, such as heated Wax affords to the Seal; or else a subsiding or hardning of the Matter, after by some kind of Fluidity it had perfectly fill'd or inclosed the figuring Vegetable or Animal

Substance, after the manner as a Statue is made of Plaister of *Paris*, or Alabaster-dust beaten and boil'd, mixed with Water and poured into a Mould.

Secondly, Next that there seems to have been some extraordinary Cause which did concur to the promoting of this Coagulation or Petrification; and that every kind of Matter is not of it self apt to coagulate into a strong Substance so hard as we find most of those Bodies to consist of.

Thirdly, That the concurrent Causes assisting towards the turning of these Substances into Stone seem to have been one of these, either some kind of fiery Exhalation arising from subterraneous Eruptions or Earthquakes; or secondly, a Saline Substance, whither working by Dissolution and Congelation,* or Crystallization, or else by Precipitation and Coagulation; or thirdly, some glutinous or bituminous Matter, which upon growing dry or setling grows hard and unites sandy Bodies together into a pretty hard Stone; or fourthly, a very long continuation of these Bodies under a great degree of Cold and Compression.

Fourthly, That Waters themselves may in tract of time be perfectly transmuted into Stone, and remain a Body of that Constitution* without being reducible by any Art yet commonly known.

Fifthly, That divers other fluid Substances have, after a long continuance at rest, settled and congealed into much more hard and permanent Substances.

Sixthly, That a great part of the Surface of the Earth hath been since the Creation transformed and made of another Nature; namely, many Parts which have been Sea are now Land, and divers other Parts are now Sea which were once a firm Land; Mountains have been turned into Plains and Plains into Mountains, and the like.

Seventhly, That divers of these kind of Transformations have been effected in these Islands of *Great Britain*; and that 'tis not improbable but that many very Inland Parts of this Island, if not all, may have been heretofore all cover'd with the Sea, and have had Fishes swimming over it.

Eighthly, That most of those Inland Places where these kinds of Stones are, or have been found, have been heretofore under the Water; and that either by the departing of the Waters to another part or side of the Earth, by the alteration of the Center of Gravity of the whole Bulk, which is not impossible; or rather by the Eruption of some kind of subterraneous Fires or Earthquakes, whereby great quantities of Earth have then been rais'd above the former Level of those Parts, the Waters have been forc'd away from the Parts they formerly cover'd, and many of those Surfaces are now raised above the Level of the Water's Surface many scores of Fathoms.

Ninthly, It seems not improbable that the tops of the highest and most considerable Mountains in the world have been under Water, and that they themselves most probably seem to have been the Effects of some very great Earthquake, such as the *Alpes* and Appennine Mountains, *Caucasus*, the Pike* of *Tenariff*, the Pike in the *Terceras*, and the like.

Tenthly, That it seems not improbable but that the greatest part of the Inequality of the Earth's Surface may have proceeded from the Subversion* and tumbling thereof by some preceding Earthquakes.

Eleventhly, That there have been many other Species of Creatures in former Ages, of which we can find none at present; and that 'tis not unlikely also but that there may be divers new kinds now which have not been from the beginning.

There are some other Conjectures of mine yet unmention'd, which are more strange than these; which I shall defer the mentioning of till some other time; because tho' I have divers Observations concurring, yet having not been able to meet with such as may answer some considerable Objections that they are liable to, I will rather at present endeavour to make probable those already mentioned by setting down some of those Observations (for it would be tedious to insert all) I have collected, both out of Authors and from my own Experience.

The first Proposition confirmed.

The first was, That these figured Bodies dispersed over the World are either the Beings themselves petrify'd or the Impressions made by those Beings. To confirm which I have diligently examin'd many hundreds of these figured Bodies, and have not found the least probability of a plastick Faculty. For first, I have found the same kind of Impression upon Substances of an exceeding differing Nature, whereas Nature in other of her works does adapt the same kind of Substance to the same Shape; the Flesh of a Horse is differing from that of a Hog or Sheep, or from the Wood of a Tree, or the like; so the Wood of Box,* for Instance, is differing from the Wood of all other Vegetables; and if the outward figure of the Plant or Animal differ, to be sure their Flesh also differs. And under the same Shape you always meet with Substances of the same kind, whereas here I have observed Stones bearing the same Figure, or rather Impression, to be of hugely differing Natures, some of Clay, some of Chalk, some of Spar*, some of Marble, some of a kind of Free-stone, some like Crystals or

Diamonds, some like Flints, others a kind of Marchasite, others a kind of Ore. Nay, in the same figur'd Substance I have found divers sorts of very differing Bodies or kinds of Stone, so that one has been made up partly of Stone, partly of Clay, and partly of Marchasite, and partly of Spar, according as the Matter chanced to be jumbled together and to fill up the Mould of the Shell.

Another Circumstance which makes this Conjecture the more probable is that the outward Surface only of the Body is form'd, and that the inward Part has nothing of Shape that can reasonably be referr'd to it; whereas we see that in all other Bodies that Nature gives a Shape to, she figures* also the internal Parts or the very Substance of it with an appropriate Shape. Thus in all kinds of Minerals, as Spars, Crystals, and divers of the precious Stones, Ores, and the like the inward Parts of them are always correspondent to the outward Shape; as in Spar, if the outward Part be shap'd into a Rhomboidical parallepiped,* the inward Part of it is shap'd in the same manner, and may be cleft out into a multitude of Bodies of the like Form and Substance.

Another Circumstance is that I have in many found the perfect Shell inclosed making a concave Impression on the Body that inclosed it, and a convex on the Body that it did inclose; which I have sometimes been able to take out intire, and found it to be both by its Substance and Shape, and reflective shining, and the like Circumstances, a real Shell of a Cockle, Periwinkle, Muscle,* or the like.

And farther, I have found in the same place divers of the same kinds of Shells, not fill'd with a Matter that was capable of taking the Impression but with a kind of sandy Substance; which lying loose within it could be easily shook out, leaving the inclosing Shell perfectly intire and empty; others I have seen which have been of black Flint, wherein the Impression has been made only of a broken Shell which stuck also into it; the other Part of the Surface of that Stone which was not within the Shell remaining only form'd like a common Flint.

And which seems to confirm this Conjecture much more than any of the former Arguments, I had this last Summer an Opportunity to observe upon the South-part of *England*, in a Clift whose Bottom the Sea wash'd, that at a good heighth in the Clift above the Surface of the Water there was a Layer, as I may call it, or Vein of Shells, which was extended in length for some Miles. Out of which Layer I digg'd out and examin'd many hundreds, and found them to be perfect Shells of Cockles, Periwinkles, Muscles, and divers other sorts of small Shell-Fishes; some of which were fill'd with the Sand with which they were mix'd; others remain'd empty, and perfectly intire. From the Sea-waters washing the under part of this Clift great quantities of it do every Year tumble or founder* down and fall into the Salt-water, which are wash'd also by several Mineral-waters issuing out at the bottom of those Clifts. Of these founder'd Parts I examined very many Parcels*, and found some of them made into a kind of harden'd Mortar or very soft Stone, which I could easily with my Foot, and even almost with my Finger, crush in Pieces; others that had lain a longer time exposed to the Vicissitudes of the rising and falling Tides I found grown into pretty hard Stones; others that had been yet longer I found converted into very hard Stone, not much yielding to the hardness of Flints. Out of divers of these I was able to break and beat out divers intire and perfect Shells, fill'd with a Substance which was converted into a very hard Stone, retaining exactly the Shape of the inclosing Shell. And in the part of the Stone which had encompass'd the Shell there was left remaining the perfect Impression and Form of the Shell; the Shell it self remained as yet of its natural white Substance, though much decay'd or rotted by time. But the Body inclosing and included by the Shell I found exactly stamp'd like those Bodies, whose Figures Authors generally affirm to be the Product of a Plastick or Vegetative Faculty working in Stones....

The fourth Proposition confirmed.

The Fourth Proposition therefore to be explain'd and made probable is That Waters themselves of divers Kinds are, and may have been transmuted perfectly into a stony Substance of a very permanent Constitution, being scarcely reducible again into Water by any Art yet commonly known. And that divers other Liquid or Fluid Substances have in tract* of time settled and congealed into much more hard, fixt, solid and permanent Forms than they were of at first.

The probability of which Proposition may appear from these Particulars.

I. That almost in all Streams and running Waters there is to be found great quantity of Sand at the bottom, many of which Sands both by their Figure in the Microscope, and transparently, seem to have been generated out of the Water.

First, I say, That their transparency which they discover in the Microscope is an Argument, because I believe there is no transparent Body in the World that has not been reduc'd to that Constitution by being some ways or other made fluid, nor can I indeed imagine how there should be any. All Bodies made transparent by Art must be reduc'd into that Form first, and therefore 'tis not unlikely but that

Nature may take the same Course; but this as only probable I shall not insist on. Next, I say, that the Figures of diverse of them in the Microscope discover the same things, for I have seen multitudes of them curiously wrought and figured like Crystal or Diamonds, and I cannot imagine by what other Instrument Nature should thus cut them, save by Crystalizing them out of a Liquid or Fluid Body, and that way we find her to work in the formation of all those curious regular Figures of Salts, and the Vitriols* (as I may call them) of Metals and divers other Bodies, of which Chymistry affords many Instances. Sea-Salt and Salgem* chrystylizeth into Cubes or four-sided Parrallelipipeds,* Niter into triangular and hexangular Prisms; Alume into Octoedrons, Vitriols into various kinds of Figures according to the various kinds of Metals dissolved, and the various *Menstrua* dissolving them; Tartars* also, and Candyings* of Vegetables are figured into their various regular Shapes from the same Method and Principle. And in truth, in the formation of any Body out of this mineral Kingdom whose Origine we are able to examine, we may find that Nature first reduces the Bodies to be wrought on into a liquid or soft Substance, and afterwards forms and shapes it into this or that Figure . . . .

The fifth Proposition proved.

Fifth Proposition will follow of consequence, *viz.* That divers other fluid Substances have, after long continuance of rest, settled and congealed into much more hard and permanent Substances. For if Water it self may be so changed and metamorphosed, which seems the farthest removed from the nature of a solid Body, certainly those which are nearer to that Nature and are mixt with such Waters will more easily be coagulated. I shall not therefore any farther insist on the Proof of this than only to mention two Particulars, and that because we have almost every where so many Instances and Experiments; and the first is that of *Pliny* in the 13th Chap. of the 35th Book of his Natural History,[23] in all which Chapter he gives us divers Instances of several kinds of Earth which, by the Sea-water and Air, converted into solid and hard Stones;

The Second is an Observation of my own, which I have often taken notice of and lately examined very diligently, which will much confirm these Histories of *Pliny* and this my present Hypothesis; and that is a Part of the Observation I have already mentioned which I made upon the Western Shore of the Isle of *Wight*. I observed a Cliff of a pretty height which, by the constant washing of the Water at the bottom of it is continually, especially after Frosts and great Rains, foundering and tumbling down into the Sea underneath it. Along the Shore underneath this Cliff are a great number of Rocks and large Stones confusedly placed, some covered, others quite out of the Water; all which Rocks I found to be compounded of Sand and Clay, and Shells, and such kind of Stones as the Shore was covered with. Examining the Hardness of some that lay as far into the Water as the Low-Water-mark I found them to be altogether as hard, if not much harder than *Portland* or *Purbeck*-stone. Others of them that lay not so far into the Sea I found much softer, as having in probability not been so long exposed to the Vicissitudes of the Tides. Others of them I found so very soft that I could easily with my Foot crush them and make Impressions into them, and could thrust a Walking-stick I had in my Hand a great depth into them. Others that had been but newly foundered down were yet more soft, as having been scarce wash'd by the Salt Water. All these were perfectly of the same Substance with the Cliff from whence they had manifestly tumbled, and consisted of Layers of Shells, Sand, Clay, Gravel, Earth, *&c.* and from all the Circumstances I could examine I do judge them to have been the Parts of the Neigbouring Cliff foundered down, and rowl'd and wash'd by degrees into the Sea, and by the petrifying Power of the Salt Water converted into perfect hard compacted Stones. I have likewise since observed the like *Phænomena* on other Shores...

The 6th Proposition confirm'd.

But to spend no more time on the proof of that of which we have almost every where Instances, divers of which I have already mention'd, I shall proceed to the 6th Proposition; which is That a great Part of the Surface of the Earth hath been since the Creation transform'd and made of another Nature: that is, many Parts which have been Sea are now Land, and others that have been Land are now Sea; many of the Mountains have been Vales, and the Vales Mountains, *&c.*

For the proving of which Proposition I shall not need to produce any other Arguments besides the repeating what I find set down by divers Natural Historians concerning the prodigious Effects that have been produced by Earthquakes on the superficial Parts of the Earth, because they seem to me to have been the chief Efficients* which have transported these petrify'd Bodies, Shells, Woods, Animal Substances, *&c.* and left them in some Parts of the Earth as are no other ways likely to have been the Places wherein such Substances should be produced; they being usually either raised a great way above the level Surface of the Earth, on the Tops of high Hills, or else buried a great way beneath that Surface in the lower Valleys. For who can imagine that Oysters, Muscles, and Periwinkles, and the like Shell-fish should ever have had their Habitation on the Tops of the Mountain *Caucasus?* Which is by divers of our Geographers accounted as high in its perpendicular Altitude as any Mountain in the yet known World; and yet *Olearius*[24] affords us a very considerable History to this purpose of his own Observation, which I shall hereafter have occasion to relate and examine more particularly. Or to come a little nearer home, who could imagine that Oysters, *Echini*, and some other Shell-fish should heretofore have lived at the tops of the *Alps*, *Appennine*, and *Pyrenian* Mountains, all which abound with great store of several sorts of Shells; nay, yet nearer, at the tops of some of the highest in *Cornwal* and *Devonshire*, where I have been informed by Persons whose Testimony I cannot in the least suspect, that they have taken up divers, and seen great Quantities of them? And to come yet nearer, who can imagine Oysters to have lived on the Tops of some Hills near *Banstead-Downs* in *Surry?* Where there have been time out of Mind, and are still to this day found divers Shells of Oysters, both on the uppermost Surface and buried likewise under the Surface of the Earth, as I was lately informed by several very worthy Persons living near those Places, and as I my self had the Opportunity to observe and collect.

The Effects of Earthquakes.

To proceed then to the Effects of Earthquakes, we find in Histories Four Sorts or *Genus's* to have been performed by them.

The first is the raising of the superficial Parts of the Earth above their former Level: and under this Head there are Four Species. The 1st is the raising of a considerable Part of a Country which before lay level with the Sea, and making it lye many Feet, nay, sometimes many Fathoms above its former height. A 2d is the raising of a considerable part of the bottom of the Sea, and making it lye above the Surface of the Water, by which means divers Islands have been generated and produced. A 3d Species is the raising of very considerable Mountains out of a plain and level Country. And a 4th Species is the raising of the Parts of the Earth by the throwing on of a great Access of new Earth, and for burying the former Surface under a covering of new Earth many Fathoms thick.

A second sort of Effects perform'd by Earthquakes is the depression or sinking of the Parts of the Earth's Surface below the former Level. Under this Head are also comprized Four distinct Species, which are directly contrary to the four last named.

The *First* is a sinking of some Part of the Surface of the Earth lying a good way within the Land, and converting it into a Lake of an almost unmeasurable depth.

The *Second* is the sinking of a considerable Part of the plain Land near the Sea below its former Level, and so suffering the Sea to come in and overflow it, being laid lower than the Surface of the next adjoining Sea.

A *Third* is the sinking of the Parts of the bottom of the Sea much lower, and creating therein vast *Vorages** and *Abysses*.

A *Fourth* is the making bare or uncovering of divers Parts of the Earth which were before a good way below the Surface, and this either by suddenly throwing away these upper Parts by some subterraneous Motion, or else by washing them away by some kind of Eruption of Waters from unusual Places, vomited out by some Earthquake.

A Third sort of Effects produced by Earthquakes are the Subversions, Conversions, and Transpositions of the Parts of the Earth.

A Fourth sort of *Effects* are *Liquefaction, Baking, Calcining**, *Petrifaction, Transformation, Sublimation, Distillation, &c.*

[Hooke gives examples of each kind of earthquake from a wide range of historical and topographical works, ancient and modern.]

The Motion of the Water another cause of Alteration on the Earth.

Another Cause there is which has been also a very great Instrument in the promoting the alterations on the Surface of the Earth, and that is the motion of the Water; whether caus'd *1st*. By its Descent from some higher place, such as Rivers and Streams caus'd by the immediate falls of Rain, or Snow, or by the melting of Snow from the sides of Hills. Or, *2dly*. By the natural Motions of the Sea, such as are the Tides and Currents. Or, *3dly*. By the accidental motions of it caus'd by Winds and Storms. Of each of these we have very many Instances in Natural Historians, and were they silent the constant Effects would daily speak as much. The former Principle seems to be that which generates Hills and Holes, Cliffs and Caverns, and all manner of Asperity* and irregularity in the Surface of the Earth; and this is that which indeavours to reduce them back again to their pristine* Regularity, by washing down the tops of Hills and filling up the bottoms of Pits, which is indeed consonant to all the other methods of Nature in working with contrary Principles of Heat and Cold,

Driness and Moisture, Light and Darkness, &c. by which there is, as it were, a continual circulation. Water is rais'd in Vapours into the Air by one Quality and precipated down in drops by another, the Rivers run into the Sea, and the Sea again supplies them. In the circular Motion of all the Planets there is a direct Motion which makes them indeavour to recede from the Sun or Center, and a magnetick or attractive Power that keeps them from receding. Generation creates and Death destroys; Winter reduces* what Summer produces. The Night refreshes what the Day has scorcht, and the Day cherishes what the Night benumb'd. The Air impregnates the Ground in one place, and is impregnated by it in another. All things almost circulate and have their Vicissitudes. We have multitudes of instances of the wasting* of the tops of Hills and of the filling or increasing of the Plains or lower Grounds, of Rivers continually carrying along with them great quantities of Sand, Mud, or other Substances from higher to lower places; Of the Seas washing Cliffs away and wasting the Shores; Of Land Floods carrying away with them all things that stand in their way, and covering those Lands with Mud which they overflow, levelling Ridges and filling Ditches. Tides and Currents in the Sea act in all probability what Floods and Rivers do at Land; and Storms effect that on the Sea Coasts that great Land Floods do on the Banks of Rivers. *Ægypt* as lying very low and yearly overflow'd is inlarg'd by the sediment of the *Nile*, especially towards that part where the *Nile* falls into the *Mediterranean*. The Gulph of *Venice* is almost choak'd with the Sand of the *Po*. The Mouth of the *Thames* is grown very shallow by the continual supply of Sand brought down with the Stream. Most part of the Cliffs that Wall in this Island do Yearly founder and tumble into the Sea. By these means many parts are covered and rais'd by Mud and Sand that lye almost level with the Water, and others are discover'd and laid open that for many Ages have been hid....

Nor are these Changes now only, but they have in all probability been of as long standing as the World. So 'tis probable there may have been several vicissitudes of changes wrought upon the same part of the Earth. It may have been of an exact spherical Form, with the rest of the Earths or Planets, at the Creation of the World, before the eternal Command of the Almighty that the Waters under the Heaven should go to their place, which before cover'd the Earth, so as that it was invisible and incompleated, and the Darkness of the Deep was over it[25] (being all over cover'd with a very thick shell of Water which environ'd it on every side, it being then in all probability created of an exact Spherical Figure, and so the Waters being of themselves lighter than the Earth must equally spread themselves over the whole Surface of the Earth), and where the Breath of the Lord moved above or upon the Surface of these Waters. It may, I say, in probability have been then a part of the exact Sphœrical Surface of the Earth, and upon the command that the Waters under the Air or Atmosphere (which seems to be denoted by στερέωμα or Firmament, for the Hebrew Word signifies an Expansum*) should be gathered together into one place, and that the dry Land should appear. It may have been by that extraordinary Earthquake (whereby the Hills and Land were rais'd in one place, and the Pits or deeper places whether* the Water was to recede and be gathered together to constitute the Sea were sunk in another) rais'd perhaps to lye on the top of a Hill or in a Plain, or sunk into the bottom of the Sea, and by the washing of Waters in motion either carried to a lower place to cover some part of the Vale, or else be cover'd by adventitious* Earth brought down upon it from some higher place; which kind of alterations were certainly very great by the Flood of *Noah*, and several other Floods we find recorded in Heathen Writers. If at least there were not somewhat of an Earthquake which might again sink those Parts which had been formerly raised, to make the dry Land appear and raise the bottom of the Sea, which had been sunk for the gathering together of the Waters (which Opinion *Seneca* ascribes to *Fabianus*).[26] His description of the Manner and Effects of a Flood is fine and very suting to my present Hypothesis. This Part being thus covered with other Earth, perhaps in the bottom of the Sea, may by some subsequent Earthquakes have since been thrown up to the top of a Hill, where those parts with which it was by the former means covered may in tract of time by the fall and washing of Waters be again uncovered and laid open to the Air, and all those Substances which had been buried for so many Ages before, and which the devouring Teeth of Time had not consumed, may be then exposed to the Light of the Day....

The last Argument to prove the sixth Proposition from the Shells, &c. found on, and in the Earth.

But to proceed to the last Argument to confirm the 6th Proposition I at first undertook to prove, namely that very many parts of the Surface of the Earth (not now to take notice of others) have been transform'd, transpos'd, and many ways alter'd since the first Creation of it. And that which to me seems the strongest and most cogent Argument of all is this, That at the tops of some of the highest Hills, and in the bottom of some of the deepest Mines, in the midst of Mountains and Quarries of Stone, &c. divers Bodies have been and daily are found that if we thoroughly examine we shall find to be real shells of Fishes, which for these following Reasons we conclude to have been at first generated by the Plastick faculty of the Soul or Life-principle of some animal, and not from the imaginary influence of the Stars or from any Plastick faculty inherent in the Earth itself so form'd; the stress* of which Argument lies in these Particulars.

First, That the Bodies there found have exactly the Form and Matter, that is, are of the same kind of Substance for all its sensible Properties, and have the same External and Internal Figure or Shape with the Shells of Animals.

Next, That it is contrary to all the other acts of Nature, that does nothing in vain but always aims at an end, to make two Bodies exactly of the same Substance and Figure, and one of them to be wholly useless or at least without any design that we can with any plausibility imagine. The Shells of Animals, to our Reason, manifestly appear to be done with the greatest Councel and Design, and with the most excellent contrivance both for the Convenience and Ornament of that Animal to which it belongs, that the particular Structure and Fabrick of that Animal was capable of. Whereas these, if they were not the Shells of Fishes will be nothing but the sportings of Nature, as some do finely fancy, or the effects of Nature idely mocking herself, which seems contrary to her Gravity. But this perhaps may not seem so cogent, tho', if it be thoroughly consider'd there is much weight in it.

Next therefore, Wherever Nature does work by peculiar forms and Substances we find that she always joins the Body so fram'd with some other peculiar Substance. Thus the Shells of Animals, whilst they are forming are join'd with the Flesh of the Animal to which they belong. Peculiar* Flowers, Leaves, and Fruit are appropriated to peculiar Roots, whereas these on the contrary are found mixt with all kind of Substances, in Stones of all kinds, in all kinds of Earth, sometimes expos'd to the open Air without any coherence* to any thing. This is at least an Argument that they were not generated in that posture* they are found; that very probably they have been heretofore distinct and disunited from the Bodies with which they are now mixt, and that they were not formed out of these very Stones or Earth, as some imagine, but deriv'd their Beings from some preceding Principle.

Fourthly, Wherever else Nature works by peculiar forms we find her always to compleat that form, and not break off abruptly. But these Shells that are found in the middle of Stones are most of them broken, very few compleat, nay I have seen many bruised and flaw'd, and the parts at a pretty distance one from another, which is an

Argument that they were not generated in the place where they are found and in that posture, but that they have been sometimes distinct and distant from those Substances, and then only placed, broken and disfigured by chance, but had a preceeding and more noble Principle to which they ow'd their Form, and by some hand of Providence were cast into such places where they were filled with such Substances as in tract of Time have condensed and hardened into Stone. This, I think, any impartial Examiner of these Bodies will easily grant to be very probable, especially if he take notice of the Circumstances I have already mention'd. Now if it be granted that there have been preceding Moulds, and that these curiously figured Stones do not owe their form to a plastick or forming Principle inherent in their Substances, why might not these be supposed Shells, as well as other Bodies of the same Shape and Substance, generated none knows how, nor can imagine for what?

Further, if these be the apish* Tricks of Nature, Why does it not imitate several other of its own Works? Why do we not dig out of Mines everlasting Vegetables, as Grass for instance, or Roses of the same Substance, Figure, Colour, Smell? *&c.* Were it not that the Shells of Fishes are made of a kind of stony Substance which is not apt to corrupt and decay. Whereas Plants and other animal Substances, even Bones, Horns, Teeth and Claws are more liable to the universal Menstruum* of Time. 'Tis probable therefore that the fixedness of their Substance has preserved them in their pristine Form, and not that a new plastick Principle has newly generated them. Besides, why should we not then doubt of all the Shells taken up by the Sea-shore or out of the Sea (if they had none when we found them), whether they ever had any Fish in them or not? Why should we not here conceipt* also a plastick Faculty distinct from that of the Life-princi-ple of some Animal; is it because this is more like a Shell than the other? That I am sure it cannot be. Is it because 'tis more obvious how a Shell should be placed there? If so, 'twould be as good Reason to doubt if an Anchor should be found at the top of a Hill, as the Poet affirms, or an Urn or Coins buried under Ground, or in the bottom of a Mine, whether it were ever an Anchor, or an Urne, or a coined Face, or made by the plastick Faculty of the Earth, than which what could be more absurd. And those Persons that will needs be so over-confident of their Omniscience of all that has been done in the World or that could be, may, if they will vouchsafe, suffer themselves to be asked a Question, Who inform'd them? Who told them where *England* was before the Flood; nay, even where it was before the *Roman* Conquest, for about four or five thousand Years, and perhaps much longer; much more where did they ever read or hear of what *Changes* and *Transpositions* there have been of the parts of it before that? What History informs us of the burying of those Trees in *Cheshire* and *Anglesy?* Who can tell when *Tenariff* was made? And yet we find that most judicious Men that have been there and well considered the form and posture of it conclude it to have been at first that way produced.

But I suppose the most confident will quickly upon examination find that there is a defect of Natural History; if therefore we are left to conjecture then that must certainly be the best that is backed with most Reason, that Clay, and Sand, and common Shells can be changed and incorporated together into Stones very hard. I have already given many instances and can produce hundreds of others, but that I think it needless, that several parts of the bottom of the Sea have been thrown up into Islands and Mountains. I have also given divers Instances, and those some of them within the Memory of Man, where 'tis not in the least to be doubted but that there may be found some Ages hence several Shells at the tops of those Hills there generated; and as little, that if Quarries of Stone should be hereafter digged in those places, there would be found Shells incorporated with them; and were they not beholding to this inquisitive and learned Age for the History of that Eruption they might as much wonder how these Shells should come there, and ascribe them to a plastick Faculty or some imaginary Influence, as plausibly as some now do. I have also shewed that Water and divers other fluid Substances may be, in tract of Time, converted into Stone and stony Substances; and so such Liquors penetrating the Pores of these Shells, and especially if they be assisted by the benumming* Steams that sometimes issue from Subterraneous Erruptions, may very much contribute to the preservation of those Shells from Corruption and crumbling to Dust under the crushing Foot of Time. Besides, that the Shells themselves are so near the Nature and Substance of Stone that they are little subject to the injuries of the Air or Weather; so that these small pyramidal Houses of Shell-Fishes seem not less lasting Monuments than those vast piles of Stones erected by the antient Inhabitants of *Egypt*, which outvye* all the more curious Fabricks of *Grecian* and *Roman* Architecture both for their Antiquity and present Continuance. Nor do they exceed the Works of Architects for lasting only, but for Ornament, for Strength, and for Convenience.

Thus much only I shall add at present, that from what I have instanced about Petrifactions and the hardning of several Substances, it seems very probable that in the beginning the Earth consisted for the most part of fluid Substances, which by degrees have setled, congealed and concreted,* and turn'd into Stones, Minerals, Mettals, Clays, Earth, &c. And that in process of time the parts of it have by degrees concreted and lost their Fluidity, and that the Earth itself doth wax old almost in the same manner as Animals and Vegetables do; that is, that the moisture of it doth by degrees decay and wast either into Air, and from thence into the Æether; or else by degrees the Parts communicating their motion to the Fluid ether either grow moveless and hard, almost in the same manner as we find the Bodies of Animals and Vegetables when they grow old in their several proportionate times, all the Parts tend and end in solidity and fixtness, the Gelly* becomes Gristles, and the Gristles a Bone, and the Bone at length a Stone; the Skin from smooth and soft grows rough and hard, the motions grow slow, and the moveable Parts and Joints grow stiff, and all the Juices decay and are deficient. The same thing happens in Trees and other Vegetables. If therefore the Parts of the Earth have formerly, in all probability, been softer, how much more powerful might Earthquakes be then in breaking, raising, overturning, and otherwise changing the superficial Parts of the Earth? Besides, 'tis not unlikely but Earthquakes might then be much more frequent before the Fuels of those subterraneous Fires were much spent. That the Parts of the Earth do continually grow harder and fixt and concrete into Stone I think no one will deny that has consider'd the Constitution of Mountains, the Layers and Veins of them, the Substances mixt with them, the Layers of the several Earths, Sands, Clays, Stones, Minerals, &c. that are met with in diging Mines and Wells, The Nature of Petrifying Waters, the shapes of Crystals, Ores, Talks, Sparrs, and most kind of precious Stones, Marbles, Flint, Chalk, and the like, every of which are by their forms sufficiently discover'd to have been formerly fluid Bodies, and whilst fluid shaped into those forms. One or two undeniable Instances I shall add of the fluidity of Flints, and that shall be that I have now by me a Flint that has so perfectly filled the Shell of an *Echinus** and inclosed it also that it has received all the impressions of the cracks of the Shells both on the Concave and Convex Part thereof, and has exactly filled all the Holes and Pores thereof, and has so perfectly received all the shape thereof as if it were nothing but Plaister of *Paris* tempered, Wax, or Sulphur that had been melted and cast on it; notwithstanding which it is a Flint so hard as to cut Glass very readily, and is of a very singular and uniform Texture; to this I might add many others of the like kind which have the impressions of these and other Shells, and yet are some Marble, some Pebbles, some Agats,* some Marchasites, some

Ores, some Crystals, &c. Some Flints I have marked with impressions as exactly as if they had been soft Wax stamp'd with a Seal. . . .

There is yet one Argument more that to me seems very good, and that is fetcht from no less distance than the Moon and the Sun by the help of Telescopes. These Bodies, as I have formerly hinted in the latter end of my *Micrography*, seem to have the same Principle of Gravity as the Earth, which, as I have there argued, seems probable from their Spherical Figure in general, and the several inequalities in particular visible by the help of Telescopes on the Surface of the Moon, and the several Smoaks, and Clouds, and Spots that appear on the Surface of the Sun; and as they have that Principle in common with the Earth so it seems to me that they are not free from the like motions with those of an Earthquake. For as to the Moon 'tis easily to be perceiv'd through a Telescope that the whole Surface of it is covered over with a multitude of small Pits or Cavities which are incompassed round with a kind of protuberant Brim, much like the Cavities or small Pits which are left in a Pot of Alabaster Dust boyled dry by the Vapours which break out of the Body of it by the heat of the Fire; and all the inequalities that appear on the Surface of that Body seem, by their form, to have been caus'd by an Eruption of the Moon, somewhat Analogous to our Earthquakes; all those Pits in the Moon being much like the Caldera* or Vent at the top of Vulcans* here on the Earth, or like those little Pits left at the top or surface of the Alabaster Dust by the natural subsiding of that Dust in the place where the Vapours generated within the Body of it break out. I need not, I think, spend time in urging Arguments to prove the sufficient powerfulness of the Cause to produce Effects as great as any I have ascribed to it, as being able to raise as great and high Mountains as those of the *Alps, Andes, Caucasus, Montes Lunæ*, &c. especially since even of late we are often informed of as great effects elsewhere, and even of the shaking and moving those vast Mountains by our latter and more debilitated* Earthquakes, tho' those Mountains are now in probability much more compacted and tenacious* by the since acquired Petrifaction than they were before their first accumulation;* and tho' 'tis not unlikely but the Fuel or Cause of the Subterraneous Fire may be much wasted and spent by preceding Conflagarations, Yet possibly there may be yet left in other Parts sufficient Mines to produce very great effects if they shall by any accident take Fire, and 'tis not impossible but that there may be some Causes that generate and renew the Fuel, as there are others that spend and consume it.

From all which Propositions, if at least they are true, will follow many others, meer* Corollaries which may be deduced from them.

Robert Hooke (1635–1702)

First, That there may have been in preceding Ages, whole Countries either swallowed up into the Earth, or sunk so low as to be drown'd by the coming in of the Sea, or divers other ways quite destroyed; as *Plato's Atlantis,* &c.

Secondly, That there may have been as many Countries new made and produced by being raised from under the Water, or from the inward or hidden Parts of the Body of the Earth, as *England.*

Thirdly, That there may have been divers Species of things wholly destroyed and annihilated, and divers others changed and varied, for since we find that there are some kinds of Animals and Vegetables peculiar to certain places and not to be found elsewhere, if such a place have been swallowed up 'tis not improbable but that those Animal Beings may have been destroyed with them, and this may be true both of aerial and aquatick Animals. For those animated* Bodies, whether Vegetables or Animals, which were naturally nourished or refresh'd by the Air would be destroy'd by the Water. And this I imagine to be the reason why we now find the Shells of divers Fishes Petrify'd in Stone, of which we have now none of the same kind. As divers of those Snake or Snail Stones, as they call them, whereof great varieties are found about *England,* and some in *Portland,* dug out of the very midst of the Quary of a prodigious bigness, one of which I have weighing near   Pound weight, being in Diameter about   Inches, which I obtain'd from the Honourable *Henry Howard* of *Norfolk.* We have Stories that there have been Giants in former Ages of the World, and 'tis not impossible but that such there may have been and that they may have been all destroyed, both they and their Country by an Earthquake, and the Poets seem to hint as much by their *Gigantomachia.*[27]

Fifthly, 'Tis not impossible but that there may have been a preceding learned Age wherein possibly as many things may have been known as are now, and perhaps many more, all the Arts cultivated and brought to the greatest Perfection, Mathematicks, Mechanicks, Literature, Musick, Opticks, *&c.* reduced* to their highest pitch, and all those annihilated, destroyed and lost by succeeding Devastations. Atomical Philosophy seems to have been better understood in some preceding time, as also the Astronomy evinc'd by *Copernicus,* the *Ægyptian* and *Chinese* Histories tell us of many thousand Years more than ever we in Europe heard of by our Writings, if their Chronology may be granted, which indeed there is great reason to question.

Sixthly, 'Tis not impossible but that this may have been the cause of a total Deluge, which may have caused a destruction of all things then living in the Air. For if Earthquakes can raise the Surface of the Earth in one place and sink it in another so as to make it uneven and rugged with Hills and Pits, it may on the contrary level those Mountains again and fill those Pits, and reduce the Body of the Earth to its primitive roundness, and then the Waters must necessarily cover all the Face of the Earth as well as it did in the beginning of the World, and by this means not only a learned Age may be wholly annihilated, and no relicks of it left, but also a great number of the Species of Animals and Plants. And 'tis not improbable but in the Flood of *Noah* the Omnipotent might make use of this means to produce that great effect which destroyed all Flesh and every living thing, save what was saved alive in the Ark.

Seventhly, 'Tis not impossible but that some of these great alterations may have alter'd also the magnetical Directions of the Earth; so that what is now under the Pole or Æquator, or any other Degree of Latitude may have formerly been under another, for since 'tis probable that divers of these parts that have such a Quality may have been transpos'd, 'tis not unlikely but that the magnetick Axis of the whole may be alter'd by it, after the same manner as we may find by experiment on a Loadstone, that the breaking off and transposing the parts of it do cause a variation of the magnetick Axis.

I could proceed to set down a great many other Corollarys that would naturally follow from these Principles if certainly proved. But this Essay I intended only as a hint or memorandum to such Gentlemen as travel, or any other inquisitive Persons, who for the future may have better opportunities of making Observations of this kind, that they may be hereby excited or at least intreated to take notice of such Phænomena as may clear* this Inquiry tho' never so seemingly mean and trivial, since it seems not improbable but that they may discover more of the preceding duration and alterations of the World than any other Observations whatsoever, and that thence may flow such instructions as may be of some of the most considerable uses of humane Life and Society, to which end all our Philosophical Studies and Inquiries tend. Ended *Sep.* 15. 1668.

Thomas Sprat

(1635–1713)

8 *History of the Royal Society* (1667)

Thomas Sprat was educated at Wadham College, Oxford, where he came into contact with John Wilkins (then Warden) and his group of scientists. Ordained in 1660, he became chaplain to George Villiers, second Duke of Buckingham (Wilkins's patron), and subsequently Dean of Westminster and Bishop of Rochester. Sprat was elected a member of the Royal Society in 1663, proposed by Wilkins, and wrote his *History* under Wilkins's guidance. It is a work of propaganda, projecting the Society's preferred image of itself, and including excerpts from papers read before that body between 1661 and 1664. The desire to establish itself as the leading British scientific institution and to link up with the established Church of England caused Sprat to include passages attacking their common enemies, nonconformists, Sectarians, and occult scientists, all of whom threatened to obscure the "light" of reason and experiment by their "enthusiasm" (irrational excitement), private language, and jargon. If unreliable as a factual history, Sprat's work is valuable for its record of the ethos and ideals of the most long-lived co-operative scientific institution.

Source: *The History of the Royal-Society of London, For the Improving of Natural Knowledge* (London, 1667).

The Preface, and Design of this Discourse

I shall here present to the World an Account of the *First Institution* of the *Royal Society* and of the *Progress* which they have already made: in hope that this Learned and Inquisitive Age will either think their Indeavours worthy of its *Assistance*, or else will be thereby provok'd to attempt some *greater Enterprise* (if any such can be found out) for the Benefit of humane life by the Advancement of *Real Knowledge*.

Perhaps this Task which I have propos'd to my self will incurr the Censure of many Judicious Men, who may think it an over-hasty and presumptuous Attempt: and may object to me that the *History* of an Assembly which begins with so great expectations ought not to have been made publique so soon, till We could have produced very many considerable *Experiments* which they had try'd, and so have given undenyable *Proofs* of the usefulness of their undertaking.

In answer to this, I can plead for my self that what I am here to say will be far from preventing the labours of others in adorning so worthy a Subject: and is *premis'd** upon no other account than as the noblest Buildings are first wont to be represented in a few *Shadows* or small *Models:* which are not intended to be equal to the Chief Structure it self but onely to shew in little by what *Materials*, with what *Charge*, and by how *many Hands* that is afterwards to be rais'd. . . .

PART I, SECT. XVI. *MODERN EXPERIMENTERS.*

The *Third* sort of *new Philosophers* have been those who have not onely disagreed from the *Antients* but have also propos'd to themselves the right course to slow and sure *Experimenting:* and have prosecuted it as far as the shortness of their own Lives, or the multiplicity of their other affairs, or the narrowness of their Fortunes have given them leave. Such as these we are to expect to be but few: for they must devest themselves of many vain conceptions, and overcome a thousand false Images which lye like monsters in their way before they can get as far as this. And of these I shall onely mention one great Man, who had the true Imagination of the whole extent of this Enterprize as it is now set on foot; and that is the *Lord Bacon*. In whose Books there are every where scattered the best arguments that can be produc'd for the defence of Experimental Philosophy; and the best directions that are needful to promote it. All which he has already adorn'd with so much Art that if my desires could have prevail'd with some excellent Friends of mine who engag'd me to this Work, there should have been no other Preface to the *History* of the *Royal Society* but some of his Writings. But methinks, in this one Man I do at once find enough occasion to admire the strength of humane Wit and to bewail the weakness of a Mortal condition. For is it not wonderful that he, who had run through all the degrees of that *profession* which usually takes up mens whole time; who had studied and practis'd and govern'd the *Common Law:* who had always liv'd in the crowd and born the greatest Burden of Civil business:[1] should yet find leisure enough for these retir'd Studies, to excel all those men who separate themselves for this very purpose? He was a Man of strong, cleer, and powerful Imaginations: his Genius was searching and inimitable: and of this I need give no other proof than his Style it self; which as for the most part it describes mens minds as well as Pictures do their Bodies, so it did his above all men living. The course of it vigorous and majestical: The Wit Bold and Familiar: The comparisons fetch'd out of the way and yet the most easie: in all expressing a soul equally skill'd in Men and Nature. All this and much more is true of him. But yet his *Philosophical Works* do shew that a single and busie hand can never grasp all this whole Design of which we treat. His Rules were admirable: yet his *History* not so faithful* as might have been wish'd in many places: he seems rather to take all that comes than to choose, and to heap rather than to register.* But I hope this accusation of mine can be no great injury to his Memory; seeing, at the same time, that I say he had not the strength of a thousand men, I do also allow him to have had as much as twenty.

PART 2, SECT. V. *A MODEL OF THEIR WHOLE DESIGN.*

I will here, in the first place, contract into few Words the whole *summe* of their *Resolutions*; which I shall often have occasion to touch upon in *parcels**. Their purpose is, in short, to make faithful *Records* of all the Works of *Nature* or *Art*, which can come within their reach: that so the present Age and posterity may be able to put a mark on the Errors which have been strengthened by long prescriptions;* to restore the Truths that have lain neglected; to push on those which are already known to more various uses; and to make the way more passable to what remains unreveal'd. This is the compass of their Design. And to accomplish this they have indeavor'd to separate the knowledge of *Nature* from the colours* of *Rhetorick*, the devices of *Fancy*, or the delightful deceit of *Fables*. They have labor'd to inlarge it from being confin'd to the custody of a few, or from servitude to private interests. They have striven to preserve it from being over-press'd by a confus'd heap of vain and useless particulars; or from being straitned* and bounded too much up by General Doctrines. They have try'd to put it into a condition of perpetual increasing by settling an inviolable correspondence between the hand and the brain. They have studi'd to make it not onely an Enterprise of one season or of some lucky opportunity but a business of time, a steddy, a lasting, a popular, an uninterrupted Work. They have attempted to free it from the Artifice, and Humors, and Passions of Sects; to render it an Instrument whereby Mankind may obtain a Dominion over *Things* and not onely over one anothers *Judgements*. And lastly, they have begun to establish these Reformations in Philosophy not so much by any solemnity of Laws or ostentation of Ceremonies as by solid Practice and examples; not by a glorious pomp of Words but by the silent, effectual, and unanswerable Arguments of real Pro-ductions.

This will more fully appear by what I am to say on these four particulars, which shall make up this part of my Relation, the *Qualifications* of their *Members;* the *manner* of their *Inquiry;* their *weekly Assemblies;* and their *way* of *Registring.*

SECT. VII. IT CONSISTS CHIEFLY OF GENTLEMEN.

But though the *Society* entertains very many men of *particular Professions* yet the farr greater Number are *Gentlemen*, free and unconfin'd. By the help of this there was hopefull Provision made against *two corruptions* of Learning, which have been long complain'd of but never remov'd. The *one*, that *Knowledge* still degenerates to consult *present profit* too soon; the *other* that *Philosophers* have bin always *Masters* & *Scholars*, some imposing & all the other submitting, and not as equal observers without dependence.

The first of these may be call'd the *marrying of Arts too soon*, and putting them to generation* before they come to be of Age; and has been the cause of much inconvenience. It weakens their strength; it makes an unhappy disproportion in their increase, while not the *best* but the *most gainfull* of them florish. But above all it diminishes that very profit for which men strive. It busies them about possessing some petty prize while Nature it self, with all its mighty Treasures, slips from them; and so they are serv'd like some foolish Guards who, while they were earnest in picking up some small Money that the Prisoner drop'd out of his Pocket, let the Prisoner himself escape, from whom they might have got a great ransom. This is easily declam'd against but most difficult to be hindred. If any caution will serve it must be this; to commit the Work to the care of such men who, by the freedom of their education, the plenty of their estates, and the usual generosity of Noble Bloud, may be well suppos'd to be most averse from such sordid considerations.

The second Error which is hereby endeavour'd to be remedied is that the Seats of Knowledg have been for the most part heretofore not *Laboratories*, as they ought to be, but onely *Scholes*,* where some have *taught* and all the rest *subscrib'd*.* The consequences of this are very mischievous. For first, as many *Learners* as there are so many hands and brains may still be reckon'd upon as useless. It being onely the *Master's* part to examine and observe, and the Disciples to submit with silence to what they conclude. But besides this, the very inequality of the Titles of *Teachers* and *Scholars* does very much suppress and tame mens Spirits; which though it should be proper for Discipline and Education yet is by no means consistent with a free Philosophical Consultation. It is undoubtedly true that scarce any man's mind is so capable of *thinking strongly* in the presence of one whom he *fears* and *reverences* as he is when that restraint is taken off....

I shall only mention one prejudice more, & that is this, That from this onely teaching and learning there does not onely follow a continuance but an increase of the yoak upon our Reasons. For those who take their opinions from others Rules are commonly stricter Imposers* upon their Scholars than their own Authors were on them, or than the first Inventors of things themselves are upon others. Whatever the cause of this be, whether the first men are made meek and gentle by their long search and by better understanding all the difficulties of Knowledg, while those that learn afterwards, onely hastily catching things in small *Systems* are soon satisfy'd before they have broken their pride, & so become more imperious; or whether it arises from hence, that the same *meanness of Soul* which made them bound their thoughts by others Precepts makes them also *insolent* to their inferiors, as we always find *cowards* the most *cruel;* or whatever other cause may be alleg'd, the observation is certain, that the *successors* are usually more positive and Tyrannical than the *beginners* of Sects.

For here especially the [critics] may doubt of two things. The first, whether the *Royal Society*, being so numerous as it is, will not in short time be diverted from its primitive* purpose, seeing there wil be scarce enough men of Philosophical* temper always found to fill it up, and then others will crowd in who have not the same bent of mind, and so the whole business will insensibly be made rather a matter of noise and pomp than of real benefit? The second, Whether their number being so large will not afright private men from imparting many profitable secrets to them, lest they should thereby become common, and so they be depriv'd of the gain which else they might be sure of if they kept them to themselvs.

SECT. VIII. *A DEFENCE OF THE LARGENESS OF THEIR NUMBER.*

To the first I shall reply That this scruple is of no force in respect of *the Age wherein we live.* For now the Genius of *Experimenting* is so much dispers'd that even in this *Nation*, if there were one or two more such *Assemblies* settled there could not be wanting able men enough to carry them on. All places and corners are now busie and warm* about this Work, and we find many Noble Rarities to be every day given in, not onely by the hands of Learned and profess'd Philosophers but from the Shops of *Mechanicks*, from the Voyages of *Merchants*, from the Ploughs of *Husbandmen*, from the Sports, the Fishponds, the

Parks, the Gardens of *Gentlemen*; the doubt therefore will onely touch *future Ages*. And even for them too we may securely promise that they will not, for a long time, be barren of a Race of Inquisitive minds when the way is now so plainly trac'd out before them; when they shall have tasted of these first Fruits and have been excited by this Example. . . . But to say no more, it is so farr from being a blemish that it is rather the excellency of this Institution that *men of various Studies* are introduc'd. For so there will be always many sincere witnesses standing by whom self-love wil not persuade to report falsly, nor heat of invention carry to swallow a deceit too soon, as having themselves no hand in the making of the Experiment but onely in the *Inspection.** So cautious ought men to be in pronouncing even upon Matters of Fact. The whole care is not to be trusted to *single* men: not to a *Company* all of *one mind*; not to *Philosophers*; not to *devout* and religious men *alone*. By all these we have been already deluded, even by those whom I last nam'd, who ought most of all to abhorr falshood; of whom yet many have multiply'd upon us infinite Stories and false Miracles without any regard to Conscience or Truth.

To the second Objection I shall briefly answer that if all the Authors* or Possessors of extraordinary inventions should conspire to conceal all that was in their power from them, yet the *Method* which they take will quickly make abundant reparation for that defect. If they cannot come at Nature in its particular *Streams* they will have it in the *Fountain*. If they could be shut out from the Closets of *Physicians* or the Work-houses of *Mechanicks*, yet with the same or with better sorts of Instruments, on more materials, by more hands, with a more rational light they would not onely restore again the old Arts but find out, perhaps, many more of farr greater importance. But I need not lay much stress upon that hope when there is no question at all, but all or the greatest part of such *Domestick Receipts** and Curiosities will soon flow into this *publick Treasure.** How few secrets have there 'been, though never so gainful, that have been long conceal'd from the whole World by their *Authors*? Were not all the least Arts of life at first private? Were not *Watches*, or *Locks*, or *Guns*, or *Printing*, or lately the *Bow-dye,** devis'd by *particular men* but soon made *common*? If neither *chance*, nor *friendship*, nor *Treachery* of servants, have brought such things out, yet we see *ostentation* alone to be every day powerful enough to do it. This desire of glory, and to be counted *Authors* prevails on all, even on many of the dark and reserv'd *Chymists* themselves, who are ever printing their greatest mysteries, though indeed they seem to do it with so much reluctancy, and with a willingness to hide still; which makes their *style* to resemble the *smoak* in which they deal. Well then, if this disposition be so *universal* why should we think that the Inventors will be only tender* and backward to the *Royal Society?* From which they will not only reap the most *solid honor* but will also receive the strongest assurances of still retaining the *greatest part of the profit?* But if all this should fail there still remains a refuge which will put this whole matter out of dispute: and that is that the *Royal Society* will be able by degrees to purchase such extraordinary inventions which are now close lock'd up in *Cabinets*, and then to bring them into one common Stock, which shall be upon all occasions expos'd to all mens use. This is a most *heroick Intention:* For by such concealments there may come very much hurt to mankind. . . .

SECT. XII. *THEIR METHOD OF INQUIRY.*

In their *Method* of *Inquiring* I will observe how they have behav'd themselves in things that might be brought within their *own Touch and Sight:* and how in those which are so remote and hard to be come by, that about them they were forc'd to trust *the reports of others.*

In the first kind I shall lay it down as their *Fundamental Law* that whenever they could possibly get to *handle* the subject, the *Experiment* was still perform'd by some of the *Members* themselves. The want of this *exactness* has very much diminish'd the credit of former *Naturalists.** It might else have seem'd strange that so many men of Wit, setting so many hands on work, being so watchful to catch up all relations,* from Woods, Fields, Mountains, Rivers, Seas, and Lands, and scattering their Pensions* so liberally, should yet be able to collect so few Observations that have been judicious or useful. But the Reason is plain; for while they thought it enough to be onely *Receivers* of others Intelligence they have either employ'd *Ignorant* searchers, who knew not how to digest or distinguish what they found; or *frivolous*, who always lov'd to come home laden, though it were but with trifles; or (which is worst of all) *crafty*, who having perceiv'd the humours* of those that paid them so well would always take care to bring in such collections as might seem to agree with the Opinions and Principles of their *Masters*, however they did with *Nature* itself.

This Inconvenience the *Royal Society* has escap'd by making the whole process pass under its own eyes. And the Task was divided amongst them by one of these two ways. First, it was sometimes referr'd to some *particular men* to make choice of what *Subject* they pleased, and to follow their own humour in the *Trial*, the *expence* being still allow'd from the general Stock. By which liberty that they afforded they had a very necessary regard to the power of *particular*

Inclinations, which in all sorts of *Knowledg* is so strong that there may be numberless Instances given of men who in some things have been altogether *useless*, and yet in others have had such a vigorous and *successful* faculty as if they had been born and form'd for them alone.

Or else secondly, the *Society* it self made the distribution, and deputed whom it thought fit for the prosecution of such or such Experiments. And this they did either by allotting the *same Work* to *several* men separated one from another, or else by *joyning* them into Committees (if we may use that word in a Philosophical sense, and so in some measure purge it from the ill sound, which it formerly had).[2] By this *union* of *eyes* and *hands* there do these *advantages* arise. Thereby there will be a full *comprehension* of the object in *all* its appearances, and so there will be a mutual communication of the light of one *Science* to another, whereas *single labours* can be but as a prospect* taken upon one side. And also by this fixing of several mens thoughts upon one thing there will be an excellent cure for that *defect* which is almost unavoidable in great *Inventors*. It is the custom of such earnest and powerful minds to do wonderful things in the *beginning*, but shortly after to be overborn by the multitude and weight of their own thoughts; then to yield and cool by little and little; and at last grow weary, and even to loath that upon which they were at first the most eager .... For this the best provision must be to join many men together ....

SECT. XVI. *THEIR DIRECTING EXPERIMENTS.*

Towards the first of these ends it has been their usual course, when they themselves appointed the *Trial*, to propose one week some particular *Experiments* to be prosecuted the next, and to debate before hand concerning all things that might conduce to the better carrying them on. In this *Preliminary Collection** it has been the custom for any of the *Society* to urge what came into their thoughts or memories concerning them, either from the observations of others, or from *Books*, or from their own *Experience*, or even from common *Fame* it self. And in performing this they did not exercise any great rigour of choosing and distinguishing between *Truths* and *Falshoods*, but amass altogether as they came the certain Works, the Opinions, the Ghesses, the Inventions, with their different Degrees and Accidents, the Probabilities, the Problems, the general Conceptions, the miraculous Stories, the ordinary Productions, the changes incident to the same Matter in several places, the Hindrances, the Benefits, of *Airs*,* or *Seasons*, or *Instruments*; and whatever they found to have been begun, to have fail'd, to have succeeded in the Matter which was then under their Disquisition.*

This is a most necessary preparation to any that resolve to make a perfect search. For they cannot but go blindly and lamely and confusedly about the business unless they have first laid before them a full *Account* of it.... It is impossible but they who will onely transcribe their own thoughts, and disdain to measure or strengthen them by the assistance of others, should be in most of their apprehensions too narrow, and obscure by setting down things for general which are onely peculiar to themselves. It cannot be avoided but they will commit many gross mistakes; and bestow much useless pains by making themselves wilfully *ignorant* of what is already known and what conceal'd....

All *Knowledg* is to be got the same way that a Language is, by *Industry*, *Use*, and *Observation*. It must be receiv'd before it can be drawn forth. 'Tis true, the mind of Man is a Glass, which is able to represent to it self all the Works of *Nature*. But it can onely shew those Figures which have been brought before it. It is no *Magical Glass*,[3] like that with which *Astrologers* use to deceive the Ignorant by making them believe that therein they may behold the Image of any *Place* or *Person* in the World, though never so farr remov'd from it. I know it may be here suggested that they who busie themselves much abroad about learning the judgments of others cannot be unprejudic'd in what they think. But it is not the *knowing* but the peremptory *addiction* to others *Tenents** that sowers* and perverts the *Understanding*. Nay, to go farther, that man who is thoroughly acquainted with all sorts of *Opinions* is very much more unlikely to adhere obstinately to any one particular than he whose head is onely fill'd with thoughts that are all of one colour.

... Well then, by this first *Comment* and *Discourse* upon the *Experiment*, he that is to try it being present, and having so good an opportunity of comparing so many other mens conceptions with his own and with the *thing* it self, must needs have his thoughts more enlarg'd, his judgment confirm'd, his eyes open'd to discern what most compendious helps may be provided, what part of it is more or less useful, and upon what side it may be best attempted. The *Truths* which he learns this way will be his Pattern; the *Errors* will be his Sea-marks,* to teach to avoid the same dangers; the very falshoods themselves will serve to enlarge, though they do not inform his *Understanding*. And indeed, a thousand more advantages will hereby come into the minds of the most Sagacious and acute *Inquirers*, which they would never have compass'd if they had been onely left to themselves....

SECT. XVII. *THEIR JUDGING OF THE MATTER OF FACT.*

Those to whom the conduct of the *Experiment* is committed, being dismiss'd with these advantages, do (as it were) carry the eyes and the imaginations of the whole company into the *Laboratory* with them. And after they have perform'd the *Trial* they bring all the *History* of its *process* back again to the *test*. Then comes in the second great Work of the *Assembly*, which is to *judg* and *resolve* upon the matter of *Fact*. In this part of their imployment they us'd* to take an exact view of the repetition of the whole course of the *Experiment*. Here they observ'd all the *chances* and the *Regularities* of the proceeding; what *Nature* does willingly, what constrain'd; what with its own power, what by the succours of Art; what in a constant rode,* and what with some kind of sport* and extravagance;* industriously marking all the various shapes into which it turns it self when it is persued, and by how many secret passages it at last obtains its end;[4] never giving it over till the whole *Company* has been fully satisfi'd of the certainty and constancy, or, on the otherside, of the absolute impossibility of the effect. This *critical* and *reiterated scrutiny* of those things which are the plain objects of their eyes must needs put out of all reasonable dispute the reality of those operations which the *Society* shall positively determine to have succeeded. If any shall still think it a just *Philosophical liberty* to be jealous* of resting on their credit they are in the right; and their *dissentings* will be most thankfully receiv'd, if they be establish'd on solid works and not onely on *prejudices* or *suspicions*. To the *Royal Society* it will be at any time almost as acceptable to be *confuted* as to *discover*, seeing by this means they will accomplish their main *Design*. Others will be inflam'd, many more will labour and so the *Truth* will be obtain'd between them, which may be as much promoted by the *contentions* of hands and eyes as it is commonly injur'd by those of Tongues. However, that men may not hence undervalue their *authority*, because they themselves are not willing to impose and to usurp a *dominion* over their *reason*, I will tell them that there is not any one thing which is now approv'd and practis'd in the World that is confirm'd by stronger evidence than this which the *Society* requires; except onely the *Holy Mysteries of our Religion*. In almost all other matters of *Belief*, of *Opinion*, or of *Science* the assurance whereby men are guided is nothing near so firm as this. And I dare appeal to all *sober men* whether, seeing in all Countreys that are govern'd by Laws, they expect no more than the consent of two or three witnesses, in matters of life and estate, they will not think they are fairly dealt withall in what concerns their *Knowledg*, if they have the concurring Testimonies of *threescore or an hundred*?

Thomas Sprat (1635–1713)

SECT. XVIII. *THEIR CONJECTURING ON THE CAUSES.*

The *History* of the *Trial* perform'd being thus secur'd,* I will next declare what room they allow'd for conjecturing upon the *Causes*; about which they also took some pains, though in a farr different way from the antient *Philosophers*, amongst whom scarce any thing else was regarded but such *general contemplations*. This indeed is the *Fatal point*, about which so many of the greatest *Wits* of all Ages have miscarried; and commonly, the greater the Wit the more has been the danger. So many wary steps ought to be troden in this uncertain path, such a multitude of pleasing *Errors*, false *Lights*, disguised *Lies*, deceitful *Fancies* must be escap'd; so much care must be taken to get into the right way at first; so much to continue in it; and at last the greatest caution still remaining to be us'd, lest when the treasure is in our view we undo all by catching at it too soon with too greedy and rash a hand. These and many more are the difficulties to be pass'd, which I have here with less apprehension reckon'd up because the remedy is so nigh. To this *Work* therefore the *Society* approaches with as much circumspection and modesty* as humane counsels are capable of. They have been cautious to shun the overweening *dogmatizing* on causes on the one hand, and not to fall into a *speculative* Scepticism* on the other; and whatever causes they have with just deliberation found to hold good they still make them increase their benefits by farther experimenting upon them, and will not permit them to rust or corrupt for want of use. If after all this they shall not seem wholly to have remov'd the *mischiefs* that attend this *hazardous matter*, they ought rather to be judg'd by what they have done towards it above others than by what they have not provided against: seeing the thing it self is of that nature that it is impossible to place the minds of men beyond all condition of erring about it.

SECT XX. *THEIR MANNER OF DISCOURSE*

Thus they have directed, judg'd, conjectur'd upon and improved *Experiments*. But lastly, in these and all other businesses that have come under their care there is one thing more about which the *Society* has been most sollicitous, and that is, the manner of their *Discourse:* which, unless they had been very watchful to keep in due temper,* the whole spirit and vigour of their *Design* had been soon eaten out by the luxury and redundance of *speech*. The ill effects of this superfluity of talking have already overwhelm'd most other *Arts* and *Professions*; insomuch that when I consider the means of *happy living* and the causes of their corruption I can hardly forbear recanting what I said before, and concluding that *eloquence* ought to be banish'd out of all *civil Societies* as a thing fatal to Peace and good Manners. To this opinion I should wholly incline if I did not find that it is a Weapon which may be as easily procur'd by *bad* men as *good*, and that if these should onely cast it away and those retain it the *naked Innocence* of vertue would be upon all occasions expos'd to the *armed Malice* of the wicked.[5] This is the chief reason that should now keep up the Ornaments of speaking in any request, since they are so much degenerated from their original usefulness. They were at first, no doubt, an admirable Instrument in the hands of *Wise Men*, when they were onely employ'd to describe *Goodness, Honesty, Obedience* in larger, fairer and more moving Images, to represent *Truth* cloth'd* with Bodies, and to bring *Knowledg* back again to our very senses, from whence it was at first deriv'd to our understandings. But now they are generally chang'd to worse uses. They make the *Fancy* disgust* the best things, if they come sound and unadorn'd; they are in open defiance against *Reason*, professing not to hold much correspondence with that but with its Slaves, *the Passions*; they give the mind a motion* too changeable and bewitching to consist with *right practice*. Who can behold without indignation how many mists and uncertainties these specious *Tropes* and *Figures* have brought on our Knowledg? How many rewards which are due to more profitable and difficult *Arts* have been still snatch'd away by the easie vanity of *fine speaking*? For now I am warm'd with this just Anger, I cannot with-hold my self from betraying the shallowness of all these seeming Mysteries upon which *we Writers* and *Speakers* look so bigg. And, in few words, I dare say that of all the Studies of men nothing may be sooner obtain'd than this vicious abundance of *Phrase*, this trick of *Metaphors*, this volubility of *Tongue* which makes so great a noise in the world. But I spend words in vain, for the evil is now so inveterate that it is hard to know whom to *blame* or where to begin to *reform*. We all value one another so much upon this beautiful deceipt, and labour so long after it in the years of our education that we cannot but ever after think kinder of it than it deserves. And indeed, in most other parts of Learning I look on it to be a thing almost utterly desperate in its cure: and I think it may be plac'd amongst those *general mischiefs*, such as the *dissention* of Christian Princes, the *want of practice* in Religion, and the like, which have been so long spoken against that men are become insensible about them, every one shifting off the fault from himself to others, and so they are only made bare common places* of complaint. It will suffice my present purpose to point out what has been done by the *Royal Society* towards the correcting of its excesses in *Natural Philosophy*, to which it is, of all others, a most profest enemy.

They have therefore been most rigorous in putting in execution the only Remedy that can be found for this *extravagance:* and that has been a constant Resolution to reject all the amplifications, digressions, and swellings* of style, to return back to the primitive purity and shortness when men deliver'd so many *things* almost in an equal number of *words*. They have exacted from all their members a close, naked, natural way of speaking; positive expressions; clear senses; a native easiness; bringing all things as near the Mathematical plainness as they can; and preferring the language of Artizans, Countrymen and Merchants, before that of Wits, or Scholars*....

SECT. XXI. *THEIR WAY OF REGISTERING.*

And now to come to a close of the second part of the *Narration*. The *Society* has reduc'd* its principal observations into one *commonstock*, and laid them up in publique *Registers*, to be nakedly transmitted to the next Generation of Men, and so from them to their Successors. And as their purpose was to heap up a mixt Mass of *Experiments*, without digesting them into any perfect model, so to this end they confin'd themselves to no order of subjects, and whatever they have recorded they have done it not as compleat Schemes of opinions but as bare unfinish'd Histories.

In the order of their *Inquisitions* they have been so free that they have sometimes committed themselves to be guided according to the seasons of the year; sometimes according to what any foreiner or English Artificer* being present has suggested; sometimes according to any extraordinary accident in the Nation, or any other casualty* which has hapned in their way. By which roving* and unsettled course, there being seldome any reference of one matter to the next, they have prevented others, nay even their own hands, from corrupting or contracting the work; they have made the raising of *Rules* and *Propositions* to be a far more difficult *task* than it would have been if their *Registers* had been more *Methodical*. Nor ought this neglect of consequence and order to be only thought to proceed from their *carelessness*, but from a mature and well grounded *præmeditation*. For it is certain that a too sudden striving to reduce the *Sciences* in their beginnings into Method, and Shape, and Beauty has very much retarded their increase. And it happens to the Invention of Arts as to children in their younger years, in whose Bodies the same *applications** that serve to make them strait, slender, and comely are often found very mischievous to their ease, their strength, and their growth.

By their fair, and equal, and submissive way of *Registring* nothing but *Histories* and *Relations* they have left room for others that shall succeed to *change*, to *augment*, to *approve*, to *contradict* them at their discretion. By this they have given *posterity* a far greater power of judging them than ever they took over those that went before them. By this they have made a firm *confederacy* between their own *present labours* and the Industry of *Future Ages*, which how beneficial it will prove hereafter we cannot better ghesse than by recollecting what wonders it would in all likelyhood have produc'd e're this if it had been begun in the Times of the *Greeks*, or *Romans*, or *Scholemen*,* nay even in the very last resurrection of learning. What depth of *Nature* could by this time have been hid from our view? What Faculty of the Soul would have been in the dark? What part of human infirmities not provided against? if our Predecessors, a thousand, nay even a hundred years ago had begun to add by little and little to the store: if they would have indeavour'd to be *Benefactors* and not *Tyrants* over our Reasons; if they would have communicated to us more of their *Works* and less of their *Wit*.

SECT. XXVIII. *THE PRESENT GENIUS OF OUR NATION.*

... The late times of *Civil War* and *confusion*, to make recompense for their infinite calamities, brought this advantage with them, that they stirr'd up mens minds from *long ease* and a *lazy rest*, and made them *active*, *industrious* and *inquisitive:* it being the usual benefit that follows upon *Tempests* and *Thunders* in the *State*, as well as in the *Skie*, that they purifie and cleer the *Air* which they disturb. But now since the *Kings* return the blindness of the former *Ages*, and the miseries of this last, are vanish'd away; now men are generally weary of the *Relicks* of *Antiquity*, and satiated with *Religious disputes*; now not only the *eyes* of men but their *hands* are open, and prepar'd to *labour*. Now there is a universal *desire* and *appetite* after *knowledge*, after the peaceable, the fruitful, the nourishing *Knowledge*, and not after that of antient Sects, which only yielded hard indigestible arguments, or sharp *contentions* instead of *food*; which when the minds of men requir'd *bread* gave them only a *stone*, and for *fish* a *serpent*.

SECT. XXIX. *THE SUBJECTS ABOUT WHICH THEY HAVE BEEN EMPLOI'D.*

Whatever they have hitherto attempted on these Principles and incouragements, it has been carry'd on with a vigorous spirit and wonderful good Fortune, from their first constitution down to this day. Yet I overhear the whispers and doubts of many who demand, what they have done all this while? and what they have produc'd that is answerable to these mighty hopes which we indeavour, to make the world conceive of their undertaking?

If those who require this Account have themselves perform'd any worthy things in this space of time, it is fit that we should give them satisfaction. But they who have done nothing at all have no reason to upbraid the *Royal Society* for not having done as much as they fancy it might. To those therefore who excite it to work by their examples, as well as words and reproofs, methinks it were a sufficient Answer if I should only repeat the particulars I have already mention'd, wherein the *King* has set on foot a *Reformation* in the Ornaments and Advantages of our Country. For though the original praise of all this is to be ascrib'd to the Genius of the *King* himself, yet it is but just that some honour should thence descend to this Assembly, whose purposes are conformable to his Majesties performances of that Nature. Seeing all the little scandals that captious* humours have taken against the *Royal Society* have not risen from their general proceedings but from a few pretended offences of some of their private Members, it is but reason that we should alledge in their commendation all the excellent Designs which are begun by the *King*, who has not only stil'd himself their *Founder* but acted as a particular *Member* of their Company.

To this I will also add that in this time they have pass'd through the first difficulties of their *Charter* and Model, and have overcome all oppositions which are wont to arise against the beginnings of great things. This certainly alone were enough to free them from all imputation of idleness, that they have fram'd such an Assembly in six years which was never yet brought about in six thousand. Besides this the world is to consider that if any shall think the whole compass of their work might have come to a sudden issue,* they seem neither to understand the intentions of the *Royal Society* nor the extent of their task. It was never their aim to make a violent dispatch*. They know that precipitancy in such matters was the fault of the Antients, And they have no mind to fall into the same error which they indeavour to correct. They began at first on so large a Bottom* that it is impossible the whole Frame* should be suddenly compleated. 'Tis true, they that have nothing else to do but to express and adorn conclusions of Knowledge already made, may bring their Arts to an end as soon as they please. But they who follow the slow and intricate method of Nature cannot have the seasons of their productions so much in their own power. If we would alwayes exact from them daily or weekly harvests we should wholly cut off the occasions of very many excellent

Inventions, whose subjects are remote and come but seldome under their consideration. If we should require them immediately to reduce all their labours to publick and conspicuous* use by this dangerous speed, we should draw them off from many of the best Foundations of Knowledge. Many of their noblest discoveries, and such as will hereafter prove most serviceable, cannot instantly be made to turn to profit. Many of their weightiest and most precious *Observations* are not alwayes fit to be expos'd to open view. For it is with the greatest Philosophers as with the richest Merchants, whose Wares of greatest bulk and price lie commonly out of sight, in their Warehouses and not in their Shops.

This being premis'd, I will however venture to lay down a brief draught of their most remarkable particulars; which may be reduc'd to these following heads: The Queries and Directions they have given abroad; the Proposals, and Recommendations they have made; the Relations* they have receiv'd; the Experiments they have try'd; the Observations they have taken; the Instruments they have invented; the Theories that have been proposed; the Discourses they have written or publish'd; the Repository and Library; and the Histories of Nature, and Arts, and Works they have collected.

SECT. XXX. *THEIR QUERIES AND DIRECTIONS.*

Their manner of gathering and dispersing *Queries* is this. First they require some of their particular Fellows to examine all Treatises and Descriptions of the Natural and Artificial productions of those Countries in which they would be inform'd. At the same time they employ others to discourse with the Seamen, Travellers, Tradesmen, and Merchants who are likely to give them the best light. Out of this united Intelligence from Men and Books they compose a Body of Questions concerning all the observable things of those places. These Papers being produc'd in their weekly Assemblies, are augmented or contracted as they see occasion. And then the Fellows themselves are wont to undertake their distribution into all Quarters, according as they have the convenience of correspondence: of this kind I will here reckon up some of the Principal, whose Particular heads are free to all that shall desire Copies of them for their Direction.

They have compos'd Queries and Directions what things are needful to be observ'd in order to the making of a Natural History in general; what are to be taken notice of towards a perfect History of the Air, and Atmosphere, and Weather; what is to be observ'd in the production, growth, advancing, or transforming of Vegetables; what particulars are requisite for collecting a compleat History of the Agriculture which is us'd in several parts of this Nation.

They have prescrib'd exact Inquiries, and given punctual* Advice for the tryal of Experiments of rarefaction, refraction, and condensation; concerning the cause, and manner of the Petrifaction of Wood; of the Loadstone; of the Parts of Anatomy that are yet imperfect; of Injections into the Blood of Animals; and Transfusing the blood of one Animal into another; of Currents; of the ebbing, and flowing of the Sea; of the kinds and manner of the feeding of Oysters; of the Wonders and Curiosities observable in deep Mines.

They have Collected and sent abroad Inquiries for the *East Indies*, for *China*, for *St. Helena*, for *Tenariff*, or any high Mountain, for *Ginny*,* for *Barbary*, and *Morocco*, for *Spain*, and *Portugal*, for *Turky*, for *France*, for *Italy*, for *Germany*, for *Hungary*, for *Transylvania*, for *Poland*, and *Sudan*, for *Iceland*, and *Greenland*. They have given Directions for Seamen in General, and for observing the Eclipses of the Moon; for observing the Eclipses of the Sun by *Mercury* in several parts of the World, and for observing the *Satellites* of *Jupiter*.

SECT. XXXI. THEIR PROPOSALS AND RECOMMENDATIONS

These are some of the most advantageous* *proposals* they have scatter'd and incourag'd in all places where their Interest prevails. In these they have recommended to many distinct and separate *Trials* those designs which some private men had begun but could not accomplish, by reason of their *charge*,* or those which they themselves have devis'd and conceiv'd capable of success, or even those of which men have hitherto seem'd to despair. Of these some are already brought to a hopeful issue; some are put in use, and thrive by the practice of the publick; and some are discover'd to be feasible which were only before thought imaginary and fantastical. This is one of the greatest *powers* of the true and unwearied *Experimenter*, that he often rescues things from the jaws of those dreadful Monsters, *Improbability* and *Impossibility*. These indeed are two frightful words to weaker minds, but by Diligent and Wisemen they are generally found to be only the excuses of *Idleness* and *Ignorance*. For the most part they lie not in the things themselves but in mens false *opinions* concerning them: they are rais'd by *opinions* but are soon abolish'd by *works*. Many *things* that were at first improbable to the minds of men are not so to their eyes; many that seem'd unpracticable to their *thoughts* are quite otherwise to their *hands*; many that are too difficult for their naked hands may be soon perform'd by the same hands if they are strengthen'd by *Instruments* and guided by *Method*; many that are unmanageable by a few hands, and a few Instruments, are easie to the joynt force of a multitude; many that fail in one *Age* may succeed by the renew'd indeavors of *another*. It is not therefore the conceit or fancy of men alone that is of sufficient authority to condemn the most unlikely things for *Impossible*, unless they have been often attempted in vain by many *Eyes*, many *Hands*, many *Instruments*, and many *Ages*.[6]

SECT. XXXV. *AN OBJECTION ANSWERED CONCERNING THE UNCERTAINTY OF EXPERIMENTS.*

As I am now passing away from their *Experiments* and *Observations*, which have been their proper and principal work, there comes before me an *Objection*, which is the more to be regarded because it is rais'd by the *Experimenters* themselves. For it is their common complaint that there is a great *nicety** and *contingency** in the making of many *Experiments*, that their success is very often various and inconstant, not only in the hands of *different* but even of the same *Triers*. From hence they suggest their fears that this continuance of *Experimenters*, of which we talk so much, will not prove so advantageous, though they shall be all equally cautious in *observing* and faithful in recording their *Discoveries*, because it is probable that the *Trials* of Future Ages will not agree with those of the present, but frequently thwart* and contradict them.

The *Objection* is strong, and material;* and I am so far from diminishing the weight of it that I am rather willing to add more to it. I confess many *Experiments* are obnoxious* to failing, either by reason of some *circumstances* which are scarce discernable till the work be over, or from the diversity of *Materials*, whereof some may be *genuine*, some *sophisticated*,* some *simple*, some *mix'd*, some *fresh*, some may have lost their *virtue*. And this is chiefly remarkable in *Chymical Operations*, wherein if the dissolvents be ill prepar'd, if the *Spirits** be too much or too little purify'd, if there be the least alteration in the degree of *Fire*, the quantity of *Matter*, or by the negligence of those that attend it, the whole course will be overthrown or chang'd from its first purpose.

But what is now to be concluded from hence? shall this *instability*, and *Casualty** of *Experiments* deter us from labouring in them at all? or should it not rather excite us to be more curious and watchful in their *process*? It is to be allow'd that such *undertakings* are wonderfully hazardous and difficult; why else does the *Royal Society* indeavour to preserve them from degenerating by so many *forewarnings*, and *rules*, and a *Method* so severe? It is granted that their *event* is often uncertain and not answerable to our expectations. But that only ought to admonish us of the indispensable necessity of a jealous* and exact *Inquiry*. If the uncertainty proceeded from a constant irregularity of *Nature* we had reason then to despair; but seeing it for the most part arises only from some defect or change in our progress we should thence learn, first to correct our own miscarriages before we cease to hope for the *success*.

Let then the *Experiment* be often renew'd. If the same kinds and proportions of *Ingredients* be us'd, and the same circumstances be punctually* observ'd, the *effect* without all question will be the same. If some little variation of any of these has made any alteration, a judicious and well practis'd *Trier* will soon be able to discern the *cause* of it; and to rectifie it upon the next repetition. If the difference of *time*, or *place*, or *matter*, or *Instruments* will not suffer the product to be just the same in all points: yet something else will result that may prove perhaps as beneficial. If we cannot alwayes arrive at the main end of our *Labours* some less unsought *Curiosities* will arise. If we cannot obtain that which shall be useful for practice there may something appear that may instruct.

It is strange that we are not able to inculcate into the minds of many men the necessity of that *distinction* of my Lord *Bacon's*, that there ought to be *Experiments* of *Light* as well as of *Fruit*.[7] It is their usual word, *What solid good will come from thence?* They are indeed to be commended for being so severe *Exactors** of *goodness*. And it were to be wish'd that they would not only exercise this vigour about *Experiments* but on their own *lives*, and *actions:* that they would still question with themselves in all that they do, what *solid good* will come from thence? But they are to know that in so large and so various an *Art* as this of *Experiments* there are many degrees of usefulness: some may serve for real and plain *benefit*, without much *delight*; some for *teaching* without apparent *profit*; some for *light* now, and for *use* hereafter; some only for *ornament* and *curiosity*. If they will persist in contemning all *Experiments* except those which bring with them immediate *gain* and a present *harvest*, they may as well cavil at the Providence of God that he has not made all the seasons of the year to be times of *mowing, reaping*, and *vintage*.

SECT. XL. *THE CONCLUSION OF THIS PART.*

I have now perform'd my *Promise*, and drawn out of the Papers of the *Society* an *Epitome* of the chief *Works* they have conceiv'd in their minds, or reduc'd* into Practice. If any shall yet think they have not usefully employ'd their time, I shall be apt to suspect that they understand not what is meant by a *diligent* and *profitable labouring* about *Nature*. There are indeed some men who will still condemn them for being idle unless they immediately profess to have found out the Squaring of the Circle, or the *Philosophers Stone*, or some other such mighty *Nothings*. But if these are not satisfied with what the *Society* has done they are only to blame the extravagance of their own Expectations. I confess I cannot boast of such pompous *Discoveries*. They promise no Wonders, nor endeavour after them. Their Progress has been equal, and firm, by Natural degrees and thorow* small things as well as great. They go leisurably on, but their slowness is not caus'd by their idleness but care. They have contriv'd in their thoughts, and couragiously begun an *Attempt* which all *Ages* had despair'd of. It is therefore fit that they alone, and not others who refuse to partake of their burden, should be Judges by what steps and what pace they ought to proceed.

Such men are then to be intreated not to interrupt their *Labors* with impertinent rebukes; they are to remember that the *Subject* of their *Studies* is as large as the *Univers*, and that in so vast an *Enterprise* many intervals and disappointments must be recon'd* upon. Though they do not behold that the *Society* has already fill'd the world with *perfect Sciences*, yet they are to be inform'd that the nature of their *Work* requir'd that they should first begin with *immethodical Collections* and *indigested Experiments* before they go on to finish and compose them into *Arts*. In which Method they may well be justified, seeing they have the *Almighty Creator* himself for an *Example*. For he at first produc'd a confus'd and scatter'd Light, and reserv'd it to be the *work* of another day to gather and fashion it into beautiful *Bodies*.

PART 3, SECT. XL. THE CONCLUSION, BEING A GENERAL RECOMMENDATION OF THIS DESIGN.

And now as I have spoken of a *Society* that prefers *Works* before *Words*, so it becomes their *History* to endeavor after real *fruits* and *effects*. I will therefore conclude by recommending again this *Undertaking* to the *English Nation*; to the *bravest People* the most *generous Design*; to the most zealous lovers of *Liberty* the surest way to randsome the minds of all mankind from *Slavery*.

The Privileges that our *Kings* Dominions enjoy for this end appear to be equal'd by no other *Country*. The men that we have now living to employ are excellently furnish'd with all manner of abilities. Their Method is already setled, and plac'd out of the reach of calumny or contradiction.

Thomas Sprat (1635–1713)

The work it self indeed is vast, and almost incomprehensible when it is consider'd in gross. But they have made it feasible and easie by distributing the burden. They have shew'n to the World this great secret, That *Philosophy* ought not only to be attended by a select company of *refin'd Spirits*. As they desire that its productions should be *vulgar*, so they also declare that they may be promoted by *vulgar hands*. They exact no extraordinary præparations of *Learning:* to have sound *Senses* and *Truth* is with them a sufficient Qualification. Here is enough business for *Minds* of all sizes.[8] And so boundless is the variety of these *Studies* that here is also enough delight to recompence the Labors of them all, from the most ordinary capacities to the highest and most searching *Wits*.

Here first they may take a plain view of all particular things, their kinds, their order, their figure, their place, their motion. And even this naked prospect cannot but fill their thoughts with much satisfaction, seing it was the first pleasure which the *Scripture* relates *God* himself to have taken at the *Creation*; and that not only once but at the end of every days work, when he saw all that he had made and approv'd it to be good. From this they may proceed to survey the difference of their Composition, their Effects, the Instruments* of their Beings and Lives, the Subtilty* and Structure, the decay and supply* of their parts; wherein how large is the space of their delight, seing the very shape of a *Mite* and the sting of a *Bee* appears so prodigious. From hence they may go to apply things together, to make them work one upon another, to imitate their productions, to help their defects, and with the Noblest duty to assist *Nature*, our common mother, in her *Operations*: From hence to all the works of mens hands, the divers *Artifices* of several *Ages*, the various Materials, the Improvement of *Trades*, the advancement of *Manufactures*. In which last alone there is to be found so great content that many Mighty Princes of the former and present Times, amidst the pleasures of *Government*, which are no doubt the highest in the World, have striven to excel in some *Manual Art*.

In this spacious field their *Observations* may wander, And in this whatever they shall meet with they may call their own. Here they will not only injoy the cold contentment of *Learning* but that which is far greater, of *Discovering*. Many things that have bin hitherto hidden will arise and expose themselves to their view. Many Methods of advancing what we have already will come in their way. Nay, even many of the lost *Rarities of Antiquity* will be hereby restor'd. Of these a great quantity has bin overwhelm'd in the ruines of *Time*, And they will sooner be retreiv'd by our laboring anew in the material Subjects whence they first arose than by our plodding everlastingly on the ancient *Writings*. Their *Inventions* may be soonest regain'd the same way by which their *Medals* and *Coins* have bin found; of which the greatest part has bin recover'd not by those who sought for them on purpose in old rubbish, but by digging up Foundations to rais new Buildings, and by plowing the Ground to sow new Seed.

This is the *Work* we propose to be incorag'd, which at once regards the discovering of new *Secrets* and the purifying and repairing all the profitable things of *Antiquity*. The Supply that is needed to finish it will neither impoverish Families nor exhaust a mighty income. So neer is Mankind to its happiness that so great an *Attempt* may be plentifully indow'd by a small part of what is spent on any one single Lust, or extravagant Vanity of the Time. So moderat is the *Society* in their desires of assistance that as much Charity as is bestow'd in *England* in one year, for the relief of particular Poverty and Diseases, were enough for ever to sustain a *Design* which indeavors to give aid against all the infirmities and wants of *human Nature*.

If now this *Enterprise* shall chance to fail for want of *Patronage* and *Revenew*, the world will not only be frustrated of their present expectations, but will have just ground to despair of any future *Labors* towards the increas of the *Practical Philosophy*. If our *Posterity* shall find that an *Institution* so vigorously begun and so strengthen'd by many signal advantages could not support itself, They will have reason in all times to conclude That the long barreness of *Knowledge* was not caus'd by the corrupt method which was taken but by the nature of the *Thing* itself. This will be the last great indeavor that will be made in this way if this shall prove ineffectual: and so we shall not only be guilty of our own *Ignorance* but of the *Errors* of all those that come after us.

But if (as I rather believe and præsage*) our *Nation* shall lay hold of this opportunity to deserve the applause of Mankind, the force of this *Example* will be irresistibly prævalent in all *Countries* round about us; the State of *Christendom* will soon obtain a new face; while this *Halcyon** *Knowledge* is breeding all *Tempests* will cease; the oppositions and contentious wranglings of *Science* falsly so call'd will soon vanish away; the peaceable calmness of mens *Judgments* will have admirable influence on their *Manners*; the sincerity of their *Understandings* will appear in their *Actions*; their *Opinions* will be less violent and dogmatical but more certain; they will only be *Gods* one to another and not *Wolves*;[9] the value of their *Arts* will be esteem'd by the *great things* they perform and not by those they speak. While the old *Philosophy* could only at the best pretend to the Portion of *Nephtali*, *to give goodly words,*[10] the New will have the Blessings of *Joseph* the yonger and the belov'd Son: *It shall be like a fruitful Bough, even a fruitful Bough by a Well, whose Branches run over the wall. It shall have the blessings of Heven above, the blessings of the deep that lies under, the blessings of the breasts and of the womb.*[11] While the Old could only bestow on us some barren Terms and Notions the New shall impart to us the uses of all the *Creatures*, and shall inrich us with all the Benefits of *Fruitfulness* and *Plenty*.

John Wilkins

(1614–1672)

9 An Essay Towards a Real Character and a Philosophical Language (1668)

John Wilkins had a successful clerical and academic career, being Warden of Wadham College, Oxford, from 1648 to 1659, when he became Master of Trinity College, Cambridge. Ousted at the Restoration (in 1656 he had married Cromwell's sister), he became Vicar of St. Lawrence Jewry in London (1662), and then Bishop of Chester (1668). He made no original contribution to science but produced several works popularizing new ideas for a wider public. In *The Discovery of a World in the Moon* (London, 1638) and *A Discourse Concerning a New Planet* (London, 1640) he introduced the new cosmology of Copernicus, Kepler and Galileo, rejected Aristotelianism, and suggested that it might be possible to fly to the moon. *Mercury, or the Secret and Swift Messenger* (London, 1641), is a treatise on codes and ciphers, while *Mathematical Magick, or the Wonders That May Be Performed by Mechanical Geometry* (London, 1648) explains the practical uses of mechanical devices, such as the balance, lever, wheel, pulley, wedge and screw.

Wilkins's greatest services to science were in bringing together groups of scientists to meet and share their knowledge, as in the remarkable circle at Oxford in the 1650s (Ward, Rooke, Wallis, Petty, Bathurst, Willis, Boyle, Wren, Hooke, and Sprat). At the famous meeting at Gresham College on 28 November 1660, when those who had attended a lecture by Wren met to discuss the founding of "a college for the promoting of physico-mathematical experimental learning", Wilkins chaired the meeting, and became acting president of the Royal Society from the beginning until a Royal candidate, Sir Robert Moray, was appointed. Wilkins was very active in the Royal Society, a member of the council, one of its two secretaries, sitting on many committees (for many years attending almost every meeting), and untiringly pursuing its welfare. His work on the "real character", however, historically important as the extreme phase of a distrust of language, was an embarrassment to the Royal Society, and found no followers.

Source: *An Essay Towards a Real Character and a Philosophical Language* (London, 1668).

John Wilkins (1614–1672)

Introduction

In the handling of that subject I have here proposed to treat of, I shall digest the things which to me seem most proper and material to be said upon this occasion into four parts, according to this following Method.

In the first Part I shall premise some things as *Præcognita** concerning such Tongues and Letters as are already in being, particularly concerning those various *defects* and *imperfections* in them which ought to be *supplyed* and *provided against* in any such Language or Character as is to be invented according to the rules of Art.

The second Part shall contein that which is the great foundation of the thing here designed, namely a regular *enumeration* and *description* of all those things and notions to which marks or names ought to be assigned according to their respective natures, which may be styled the *Scientifical** Part, comprehending *Universal Philosophy*. It being the proper end and design of the several branches of Philosophy to reduce all things and notions unto such a frame as may express their natural order, dependence, and relations.

The third Part shall treat concerning such helps and Instruments as are requisite for the framing of these more simple notions into continued Speech or Discourse, which may therefore be stiled the Organical or *Instrumental* Part, and doth comprehend the Art of Natural or *Philosophical Grammar*.

In the fourth Part I shall shew how these more generall Rules may be applyed to particular kinds of Characters and Languages, giving an instance of each. To which shall be adjoyned by way of *Appendix* a Discourse shewing the advantage of such a kind of Philosophical Character and Language above any of those which are now known, more particularly above that which is of most general use in these parts of the World; namely, the *Latine*.

[from Book I, ch. 3, § 5: *'Of Real Characters'*.]

Besides these, there have been some other proposals and attempts about a *Real universal Character*, that should not signifie *words* but *things* and *notions*, and consequently might be legible by any Nation in their own Tongue; which is the principal design of this Treatise. That such a Real Character is possible, and hath been reckoned by Learned men amongst the *Desiderata*, were easie to make out by abundance of Testimonies. To this purpose is that which *Piso*[1] mentions to be somewhere the wish of *Galen*, That some way might be found out to represent things by such peculiar *signs* and *names* as should express their *natures*. There are several other passages to this purpose in the Learned *Verulam*, in *Vossius*;[2] in *Hermannus Hugo*, &c. besides what is commonly reported of the men of *China*, who do now, and have for many Ages used such a general Character, by which the Inhabitants of that large Kingdom, many of them of different Tongues, do communicate with one another, every one understanding this common Character and reading it in his own Language.

It cannot be denied but that the *variety* of *Letters* is an appendix to the Curse of *Babel*, namely, the multitude and variety of *Languages*. And therefore, for any man to go about to add to their number will be but like the inventing of a Disease, for which he can expect but little thanks from the world. But this Consideration ought to be no discouragement, For supposing such a thing as is here proposed could be well established it would be the surest remedy that could be against the Curse of the Confusion,* by rendring all other *Languages* and *Characters* useless....

[I, 4, 1: '*The Defects in the Common Alphabets*'.]

One special Circumstance which adds to the Curse of *Babel* is that *difficulty* which there is in all *Languages*, arising from the various *Imperfections* belonging to them both in respect of 1. their first *Elements* or *Alphabets*, 2. their *Words*.

1. For *Alphabets*, they are all of them, in many respects, liable to just exception.*

1. As to the *Order* of them, they are inartificial and *confused*, without any such methodical distribution as were requisite for their particular natures and differences; the *Vowels* and *Consonants* being promiscuously huddled together, without any distinction. Whereas in a regular *Alphabet*, the *Vowels* and *Consonants* should be reduced into *Classes* according to their several kinds, with such an order of precedence and subsequence as their natures will bear; this being the proper end and design of that which we call *Method*, to separate the Heterogeneous, and put the Homogeneous together according to some rule of precedency.

[6: '*The Imperfections belonging to the Words of Language, as to their Equivocalness, Variety of Synonymous Words, uncertain Phraseologies, improper Way of Writing*'.]

Besides these Defects in the usual *Alphabets* or *Letters*, there are several others likewise in the *Words* of Language, and their Accidents and Constructions.

1. In regard of *Equivocals*, which are of several significations and therefore must needs render speech doubtful and obscure; and that argues a *deficiency* or want of a sufficient number of *words*. These are either *absolutely* so, or in their *figurative* construction, or by reason of *Phraseologies*.

Of the first kind there are great variety in *Latin*. So the word

	Literatos			*Codicem.*
LIBER *apud*	*Politicos*	} *significat*		*Libertate fruentem.*
	Oratores			*Filium.*
	Rusticos			*Arboris corticem.*

So the word *Malus* signifies both an *Apple-tree*, and *Evil*, and *the Mast of a ship*; and *Populus* signifies both a *Poplar-tree*, and the *People*, &c. Besides such Equivocals as are made by the *inflexion* of words: as *Lex, legis, legi; Lego, legis, legi; Sus, suis; Suo, suis; Suus, suis: Amarè* the Adverb; *Amo, amas, amavi, amare*; and *Amor, amaris vel amare*: with abundance of the like of each kind.

Nor is it better with the *English* Tongue in this respect, in which there is great variety of Equivocals. So the word *Bill* signifies both a *Weapon*, a Bird's *Beak*, and a written *Scroul:* The word *Grave* signifies both *Sober*, and *Sepulcher*, and to *Carve*, &c.

As for the ambiguity of words by reason of *Metaphor* and *Phraseology*, this is in all instituted Languages so obvious and so various that it is needless to give any instances of it; every Language having some peculiar phrases belonging to it which, if they were to be translated *verbatim* in another tongue, would seem wild and insignificant. In which our English doth too much abound, witness those words of *Break, Bring, Cast, Cleare, Come, Cut, Draw, Fall, Hand, Keep, Lay, make, Pass, Put, Run, Set, Stand, Take*, none of which have less than thirty or forty, and some of them about a hundred several senses according to their use in Phrases, as may be seen in the Dictionary. And though the varieties of Phrases in Language may seem to contribute to the elegance and ornament of Speech; yet, like other affected ornaments they prejudice* the native simplicity of it, and contribute to the disguising of it with false appearances. Besides that, like other things of fashion, they are very changeable, every generation producing new ones; witness the present Age, especially the late times, wherein this grand imposture of Phrases hath almost eaten out solid Knowledge in all professions; such men generally being of most esteem who are skilled in these Canting* forms of speech, though in nothing else.

2. In respect of *Synonymous* words, which make Language tedious and are generally *superfluities*, since the end and use of Speech is for

humane utility and mutual converse. And yet there is no particular Language but what is very obnoxious* in this kind. 'Tis said that the *Arabic* hath above a thousand several names for a *Sword*, and 500 for a *Lion*, and 200 for a *Serpent*, and fourscore for *Hony*. And though perhaps no other Language do exceed at this rate as to any particular, yet do they all of them abound more than enough in the general. The examples of this kind, for our *English*, may be seen in the following Tables. To this may be added that there are in most Languages several words that are mere *Expletives*, not adding any thing to the Sense.

3. For the *Anomalisms* and Irregularities in Grammatical construction, which abound in every Language, and in some of them are so numerous that Learned men have scrupled whether there be any such thing as *Analogy*.

4. For that *Difference* which there is in very many words betwixt the *writing* and *pronouncing* of them, mentioned before. *Scriptio est vocum pictura:*[3] And it should seem very reasonable that men should either speak as they write, or write as they speak. And yet Custom hath so rivetted this incongruity and imperfection in all Languages that it were an hopeless attempt for any man to go about to repair and amend it. 'Tis needless to give instances of this, there being in divers Languages as many words whose sounds do disagree with their way of writing as those are that agree. What is said of our *English* Tongue is proportionably true of most other Languages, That if ten Scribes (not acquainted with the particular Speech) should set themselves to write according to pronunciation, not any two of them would agree in the same way of spelling.

[I, 5, 1: *'That neither Letters nor Languages have been regularly established by the rules of Art'.*]

From what hath been already said it may appear that there are no Letters or Languages that have been at once invented and established according to the Rules of Art; but that all, except the first (of which we know nothing so certain as that it was not made by human Art upon Experience) have been either taken up from that first and derived by way of *Imitation*; or else in a long tract of time have, upon several emergencies, admitted various and *casual alterations*; by which means they must needs be liable to manifold defects and imperfections that in a Language at once invented and according to the *rules of Art* might be easily avoided. Nor could this otherwise be, because that very Art by which Language should be regulated, viz. *Grammar*, is of much *later* invention *than Languages themselves*, being adapted to what was already in being rather than the Rule of making it so....

These being some of the Defects or Imperfections in those Letters or Languages which are already known, may afford direction what is to be avoided by those who propose to themselves the Invention of a new *Character* or *Language*, which being the principal end of this Discourse I shall in the next place proceed to lay down the first Foundations of it.

[2: *The natural Ground or Principle of the several ways of Communication amongst men*.]

As men do generally agree in the same Principle of Reason so do they likewise agree in the same *Internal Notion* or *Apprehension of things*.

The *External Expression* of these Mental notions, whereby men communicate their thoughts to one another, is either to the *Ear* or to the *Eye*.

To the *Ear* by *Sounds*, and more particularly by Articulate *Voice* and *Words*.

To the *Eye* by any thing that is *visible*, Motion, Light, Colour, Figure; and more particularly by *Writing*.

That *conceit* which men have in their minds concerning a Horse or Tree is the Notion or *mental Image* of that Beast or natural thing, of such a nature, shape and use. The *Names* given to these in several Languages are such arbitrary *sounds* or *words* as Nations of men have agreed upon, either casually or designedly, to express their Mental notions of them. The *Written word* is the figure or picture of that Sound.

So that if men should generally consent upon the same way or manner of *Expression* as they do agree in the same *Notion*, we should then be freed from that Curse in the Confusion of Tongues, with all the unhappy consequences of it.

Now this can only be done either by *enjoyning* some one Language and Character to be universally learnt and practised, (which is not to be expected till some person attain to the *Universal Monarchy*, and perhaps would not be done then;) or else by *proposing* some such way as, by its facility and usefulness (without the imposition of Authority) might *invite* and ingage men to the learning of it; which is the thing here attempted.

[3: *The first thing to be provided for in the establishing of a Philosophical Character or Language is a just enumeration of all such things and notions to which names are to be assigned*.]

In order to this, The first thing to be considered and enquired into is, Concerning a just *Enumeration* and description of such things or notions as are to have *Marks* or *Names* assigned to them.

The chief Difficulty and Labour will be so to contrive the Enumeration of things and notions as that they may be full and *adæquate*, without any *Redundancy* or *Deficiency* as to the Number of them, and *regular* as to their Place and Order.

If to every thing and notion there were assigned a distinct *Mark*, together with some *provision* to express Grammatical *Derivations* and *Inflexions*, this might suffice as to one great end of a *Real Character*, namely the expression of our Conceptions by *Marks* which should signifie *things* and not *words*. And so likewise if several distinct *words* were assigned for the *names* of such things, with certain invariable *Rules* for all such Grammatical *Derivations* and *Inflexions*, and such onely as are natural and necessary; this would make a much more easie and convenient Language than is yet in being.

But now if these *Marks* or *Notes* could be so contrived as to have such a *dependance* upon, and relation to, one another as might be sutable to the nature of the things and notions which they represented; and so likewise, if the *Names* of things could be so ordered as to contain such a kind of *affinity* or *opposition* in their letters and sounds as might be some way answerable to the nature of the things which they signified; this would yet be a farther advantage superadded. By which, besides the best way of helping the *Memory* by natural Method, the *Understanding* likewise would be highly improved; and we should, by learning the *Character* and the *Names* of things, be instructed likewise in their *Natures*, the knowledg of both which ought to be conjoyned.

For the accurate effecting of this it would be necessary that the *Theory* it self, upon which such a design were to be founded, should be exactly *suted to the nature of things*. But upon supposal that this Theory is *defective*, either as to the *Fulness* or the *Order* of it, this must needs add much *perplexity* to any such Attempt and render it *imperfect*. And that this is the case with that common Theory already received need not much be doubted; which may afford some excuse as to several of those things which may seem to be less conveniently disposed of in the following Tables or Schemes proposed in the next part.

[II, 12, 1: *'A general Explication of the design of the foregoing Tables'*.]

The principal design aimed at in these Tables is to give a sufficient enumeration of all such things and notions as are to have names assigned to them, and withall so to contrive these as to their order that the place of every thing may contribute to a description of the nature of it. Denoting both the *General* and the *Particular head* under which it is placed; and the *Common difference* whereby it is distinguished from other things of the same kind.

It would indeed be much more convenient and advantageous if these Tables could be so contrived that every *difference* amongst the *Predicaments** might have a transcendental* denomination, and not depend at all upon a numerical institution. But I much doubt whether that Theory of things already received will admit of it; nor doth Language afford convenient terms by which to express several differences.

It were likewise desirable to a perfect definition of each species that the *immediate form* which gives the particular essence to every thing might be expressed; but this form being a thing which men do not know, it cannot be expected that it should be described. And therefore in the stead of it there is reason why men should be content with such a description by *properties* and *circumstances* as may be sufficient to determine the primary sense of the thing defined....

[III, 1, 1: '*Concerning the Several Kinds and Parts of Grammar*'.]

Having now dispatched the second thing proposed to be treated of, namely the *Scientifical* part, containing a regular enumeration and description of such things and notions as are to be known and to which names are to be assigned, which may be stiled *Universal Philosophy*; I proceed in the next place to the *Organical* part, or an enquiry after such kind of necessary helps, whereby as by instruments we must be assisted in the forming these more simple notions into complex Propositions and Discourses, which may be stiled *Grammar*, containing the Art of Words or Discourse.

Grammar may be distinguisht into two kinds; 1. *Natural* and *General*; 2. *Instituted* and *Particular*.

1. *Natural* Grammar (which may likewise be stiled Philosophical, Rational, and Universal) should contain all such Grounds and Rules as do naturally and necessarily belong to the Philosophy of letters and speech in the *General*.

2. *Instituted* and *Particular* Grammar doth deliver the rules which are proper and peculiar to any one Language in Particular; as about

the Inflexion of words and the Government of cases, *&c.* In the *Latin*, *Greek*, &c. and is defined by *Scaliger*[4] to be *scientia loquendi ex usu.*[5] The first of these only is upon this occasion to be considered. It hath been treated of but by few, which makes our Learned *Verulam* put it among his *Desiderata*;[6] I do not know any more that have purposely written of it.... But to me it seems that all these Authors in some measure (though some more than others) were so far prejudiced by the common Theory of the languages they were acquainted with that they did not sufficiently abstract their rules according to Nature. In which I do not hope that this which is now to be delivered can be faultless; it being very hard (if not impossible) wholly to escape such prejudices: yet I am apt to think it less erroneous in this respect than the rest....

III, 7, 1: '*Instances of the great usefulness of these Transcendental Particles, with directions how they are to be applyed*'.]

For the better explaining of what great use and advantage these Particles may be to Language, I shall give some examples severally to each pair of them according to the order premised: Beginning with the first.

I.

1.{ METAPHORICAL
    LIKE

These two are paired together because of their affinity, each of them denoting an enlargement of the sense of the word; the first *more general*, the other with reference to *Similitude*, properly so called.

The note of *Metaphorical* affixed to any Character, will signifie the enlarging the sense of that word from that strict restrained acception which it had in the Tables, to a more universal comprehensive signification. By this, common Metaphors may be legitimated, retaining their elegancy and being freed from their ambiguity. So

These words, with this note *will signifie*		These words, with this note *will signifie*	
Element	Rudiment, Principle	Shining	Illustrious
Root	Original	Hypocritical	Counterfeit
Way	Means	Banish	Expel
{ Thick / Thin	{ Gross / Subtle	Companying	Being together
{ Streight / Crooked	{ Upright / Perverse	Strengthen	Fortifie, fence
{ Obtuse / Acute	{ Dull / Quick	Wrigle in	Insinuate
{ Ripe / Immature	{ Perfect / Imperfect	Prophesie	Prediction
{ Fertile / Barren	{ Fruitful as to invention, &c. / Unfruitful	Consecrate	Dedicate
{ Beautiful / Deformed	{ Decent, Comely / Absurd, Indecent	Suiter	Candidate
{ Ornate / Homely	{ Elegant, Quaint / Rude	Woo	Canvase
{ Light / Dark	{ Evident, Plain / Mystical, Obscure	Rival	Competitor
		Raise	Prefer, Advance

So in the Tables of *Action* those Acts which are primarily ascribed unto God, as *Preserving, Destroying, Delivering, Forsaking, Blessing, Cursing,* &c. because they may by analogy be applyed to other things, therefore this mark will enlarge their acception. . . .

[IV, 1, 1: '*The Proposal of one kind of Real Character.*']

The next Enquiry should be, what kind of *Character* or Language may be fixed upon as most convenient for the expression of all those Particulars above mentioned belonging to the Philosophy of *Speech*. In order to which it may seem that the first Enquiry should be concerning *Language*; because *Writing* is but the figure of *Articulate sound* and therefore subsequent to it. But though it be true that men did first *speak* before they did *write*, and consequently *writing* is but the figure of *Speech* and therefore in order of *time* subsequent to it; yet in order of *Nature* there is no priority between these: But *voice* and *sounds* may be as well assigned to Figure* as *Figures* may be to *sounds*. And I do the rather begin with treating concerning a common *Character* or *Letter* because this will conduce more to that great end of *Facility*, whereby (as I first proposed) men are to be invited to the Learning of it. To proceed from the *Language* to the Character would require the learning of both; which being of greater difficulty than to learn one alone, is not therefore so sutable to that intention of ingaging men by the *Facility* of it. And because men that do retain their several Tongues may yet communicate by a *Real Character*, which shall be legible in all Languages; therefore I conceive it most proper to treat of this in the first place, and shall afterwards shew how this Character may be made *effable** in a distinct Language.

All Characters signifie either *Naturally* or by *Institution.** *Natural Characters* are either the Pictures of things or some other *Symbolical* Representations of them, the framing and applying of which, though it were in some degree feasible as to the general *kinds* of things, yet in most of the *particular species* it would be very *difficult*, and in some perhaps *impossible*. It were exceeding desirable that the *Names* of things might consist of such *Sounds* as should bear in them some Analogy to their *Natures*; and the Figure or Character of these Names should bear some proper resemblance to those *Sounds*, that men might easily guess at the sence or meaning of any name or word upon the first *hearing* or *sight* of it. But how this can be done in all the particular species of things I understand not; and therefore shall take it for granted that this Character must be by *Institution*. In the framing of which there are these four properties to be endeavoured after.

1. They should be most simple and easie for the Figure,* to be described by one *Ductus** of the pen or at the most by two.

2. They must be sufficiently distinguishable from one another to prevent mistake.

3. They ought to be comely and graceful for the shape of them to the eye.

4. They should be *Methodical*, Those of the same common nature having some kind of sutableness and correspondence with one another; All which qualifications would be very advantageous, both for *Understanding*, *Memory* and *Use*....

[IV, 2: *'Instances of this Real Character in the Lords Prayer'*.]

For the better explaining of what hath been before delivered concerning a Real Character it will be necessary to give some Example and Instance of it, which I shall do in the *Lords Prayer* and the *Creed*: First setting each of them down after such a manner as they are ordinarily to be written. Then the Characters at a greater distance from one another, for the more convenient figuring and interlining of them. And lastly, a Particular Explication of each Character out of the Philosophical Tables, with a Verbal Interpretation of them in the Margin.

The Lords Prayer.

[Philosophical Language characters]

Our Parent who art in Heaven, Thy Name be Hallowed, Thy

Kingdome come, Thy Will be done, so in Earth as in Heaven, Give

to us on this day our bread expedient and forgive us our trespasses as

we forgive them who trespass against us, and lead us not into

temptation, but deliver us from evil, for the Kingdome and the

Power and the Glory is thine, for ever and ever, Amen. So be it.

[IV, 4: 'An Instance of the Philosophical Language, both
in the Lords Prayer and the Creed. A Comparison of the
Language here proposed, with fifty others, as to the Facility
and Euphonicalness of it.']

As I have before given Instances of the Real Character, so I shall here
in the like method set down the same Instances for the Philosophical
Language. I shall be more brief in the particular explication of each
Word; because that was sufficiently done before in treating concerning the Character.

The Lords Prayer.

Haɪ coba ୪୪ ɪa ril dad, ha baɩbɪ ɪo ſ୪ymtɑ, ha ſalba ɪo velcɑ, ha
talbɪ ɪo vemg୪, m୪ ril dady me rɪl dad ɪo velpɪ rɑl ɑi ril ɪ poto haɪ
ſaba vaty, na ɪo ſ୪eldy୪ɪ lɑl ɑɪ haɪ balgas me ɑɪ ɪa ſ୪eldy୪ɪ lɑl
eɪ ୪୪ ɪɑ valgas r୪ ɑɪ na mɪ ɪo velco ɑɪ, rɑl bedodl୪ nil ɪo c୪ɑlbo
ɑɪ lal vɑgaɪɪe, nor ɑl ſalba, na ɑl tado, na ɑl tadalɑ ɪa ha pi୪by୪
ꟼJ m୪ ɪo.

Haɪ   coba ୪୪ ɪa ril   dad,   ha   baɩbɪ   ɪo ſ୪ymtɑ   ha
Our Father who art in Heaven, Thy Name be Hallow'ed, Thy

ſalba ɪo velcɑ, ha talbi ɪo vemg୪, m୪ ril dady me ril dad, ɪo velpɪ
Kingdome come, Thy Will be done, ſo in Earth as in Heaven, Give

rɑl ɑɪ ril ɪpoto haɪ ſaba   vaty,   na ɪo ſ୪eldi୪ɪ lal aɪ haɪ balgaɪ
to us on this day our bread expedient and forgive to us our treſpaſſes

me ɑɪ ɪa ſ୪eldy୪ɪ lal eɪ ୪୪ ɪɑ valgas r୪   ɑɪ, na mɪ ɪo velco aɪ ' rɑl
as we forgive them who treſpaſs againſt usɪ and lead us not into

bedodl୪ nil ɪo c୪ɑlbo aɪ lal vɑgaɪɪe nor ɑl ſalba, na ɑl tado, nɪ
temptation but   deliver us from evil   for the Kingdom, & the power, and

ɑl tadalɑ ɪo ha   pi୪by୪   ꟼJ m୪ ɪo.
the Glory is thine, for ever and ever.   Amen.So be it.

[IV, 6: *The Appendix, containing a comparison betwixt
this Natural Philosophical Grammar and that of other
instituted Languages, particularly the Latin, in respect of
the multitude of unnecessary Rules and of Anomalisms.*]

Having thus briefly laid the Foundations of a *Philosophical Grammar* I
am in the next place to shew the many great advantages both for
significancy, perspicuity, brevity, and consequently *facility,* which a

Character or Language founded upon these Rules must needs have above any other way of communication now commonly known or used. And because the *Latin* doth in these parts of the world supply the place of a Common Tongue, therefore I shall chiefly insist upon the comparison with that....

These things being premised concerning the many *needless Rules* and great variety of *exceptions* in the Latin, it will not be very difficult to make a comparison betwixt that and the Character and Language here proposed.

For the right estimating of the difficulty which there is in the Learning of any Language these two things are to be enquired into. 1. The *Multitude of words*. And 2. The *Grammatical Rules* belonging to such a Language.

1. As to the first of these, *Hermannus Hugo* asserts that no Language hath so few as 100,000 words; and *Varro*[7] is frequently quoted by divers Learned Men as if he affirmed that there are in the *Latin* no less than *five hundred thousand*. But upon enquiry into the scope of that place they relate to it will appear that he doth not there design to give an account of the just number of words in the *Latin*, but only to shew the great variety which is made by the *Inflexion* and *Composition* of *Verbs*. To which purpose the first thing he lays down is That there are about one thousand Radical verbs in the *Latin*. And then Secondly, That every Verb in the Declensions of it hath about five hundred several varieties or Cases of Inflexion, which make up the number of five hundred thousand. And then Thirdly, He supposeth each of these to be compounded with nine Prepositions, as for instance, the word *Cessit, Recessit, Accessit, Abscessit, Incessit, Excessit, Successit, Decessit, Concessit, Processit*; this will raise the whole number to *five millions*. In which account he reckons only the Cases and Compositions of Verbs and takes no notice of the Particles of speech, nor such other words as are not radically Verbs, which are very numerous.

Of all other Languages the *Greek* is looked upon to be one of the most copious; the Radixes of which are esteemed* to be about 3,244. But then it doth exceedingly abound in *Composition*,* in which the *Latin* Tongue being more sparing must therefore upon that account have more Radicals. What the particular number of these may be is not easie to determine; because Learned Men do not agree about many of them, whether they are *Radicals* or *Derivatives*. They may be by moderate computation estimated to be about *ten thousand*, most of which are either *absolutely*, or in *phrase*, or *both ways* equivocal ....

Many of them have no less than twenty distinct significations, and some more. Now for every several sense, we may justly reckon so many several words, which will much augment the former number. But suppose them only to treble it, and then the *Latin words* are to be reckoned thirty thousand.

2. Now for the *Latin Grammar*, it doth in the common way of Teaching take up several of our first years, not without great toyl and vexation of the mind under the hard tyranny of the School, before we arrive to a tolerable skill in it. And this is chiefly occasioned from that great multitude of such Rules as are not necessary to the Philosophy of speech, together with the *Anomalisms* and exceptions that belong to them, the difficulty of which may well be computed equal to the pains of Learning one third part of the words; according to which the labour required to the attaining of the *Latin* may be estimated equal to the pains of Learning forty thousand words.

Now in the way here proposed the words necessary for communication are not three thousand, and those so ordered by the help of natural method that they may be more easily learned and remembered than a thousand words otherwise disposed of; upon which account they may be reckoned but as one thousand. And as for such Rules as are natural to Grammar, they were not charged in the former account, and therefore are not to be allowed for here.

So that by this it appears that in point of easiness betwixt this and the Latin there is the proportion of one to forty; that is, a man of an ordinary capacity may more easily learn to express himself this way in one Month than he can by the Latin in forty Months.

This I take to be a kind of Demonstration *à Priori*; and for an Argument *à Posteriori*, namely from Experiment, though I have not as yet had opportunity of making any tryals, yet I doubt not but that one of a good Capacity and Memory may in one Months space attain to a good readiness of expressing his mind this way, either in the *Character* or *Language*.

Isaac Newton

(1642–1727)

10 *A New Theory about Light and Colours* (1672)

Isaac Newton was educated at Trinity College, Cambridge, becoming a fellow in 1667, and being appointed Lucasian professor of mathematics in 1669. He was master of the mint from 1696 to 1701, and President of the Royal Society from 1703 until his death. Knighted in 1705, he was twice elected MP for Cambridge University.

Newton's achievements in optics, physics, and mathematics made him the undisputed master of science for more than two centuries. Many of the ideas in his major work, the *Philosophiae naturalis principia mathematica* (London, 1687; rev. Cambridge, 1713, and London, 1726) were conceived in the amazingly fruitful period 1665–6, but not properly developed until after his discussions with Hooke in 1679–80, and with Edmond Halley in 1684. It was immediately recognized as a revolutionary step forward in celestial mechanics, and earned him world-wide fame.

Newton's first publications were on optics, in the *Transactions* that Oldenburg edited for the Royal Society. In 1669 Newton submitted a new reflecting telescope that he had invented, which was received with great acclaim, and won him election as fellow on 11 January 1672. Writing to Oldenburg on 18 January Newton offered to communicate "an account of a Philosophicall discovery which induced mee to the making of the said Telescope, & which I doubt not but will prove much more gratefull than the communication of that Instrument, being in my Judgment the oddest if not the most considerable detection which hath hitherto beene made in the operations of Nature" (*Correspondence*, I. 82–3). Whereas, since the time of the Greeks, white light had been thought to be pure, other colours being deviations, Newton showed that "Light itself is a *Heterogeneous mixture of differently refrangible Rays*", and that white light was in fact the compound of all the other colours. This reversal of a two-thousand-year tradition convinced some, but not all, of his contemporaries, and involved Newton in a controversy lasting several years, which caused him to shun further publication on this topic. He did not issue his *Opticks* until 1704, after Hooke's death, but enlarged it in further Latin (London, 1706) and English editions (London, 1717 or 1718), adding many "Queries" in which he published some far-reaching speculations about the nature of matter.

The form in which he announced his discovery was the new genre of the written-up experiment, which passed from Bacon to Boyle and became the standard pattern of the *Transactions* (Oldenburg slightly edited it for publication). Unlike Kepler, who had recorded every step – including the wrong ones – in his theory and calculations, Newton presents it as a wholly consistent rational experiment, successfully carried out. A problem is posed, an experiment devised that solves the problem, which leads to further problems and reflections, or confirms a scientific law. In fact, we now know that Newton worked at this topic from at least 1664 onwards, and must have met many setbacks. Rather than this example of smoothly-operating Baconian induction, Newton may have worked by profound speculation and arranged his experiment to illustrate the results of his thinking. Although Bacon had little or no influence on the content of Newton's science, Newton inherited the universal Baconian spirit distrusting science based on mere hypothesis, and not springing from direct observation of nature. In Query 31 of the *Opticks* he wrote that in both mathematics and natural philosophy analysis should always precede synthesis, and that "Analysis consists in making Experiments and Observations, and in drawing general Conclusions from them by Induction, and admitting of no Objections against the Conclusions but such as are taken from Experiments, or other certain Truths."

Source: *Philosophical Transactions*, vol. VI, no. 80 (19 February 1672), pp. 3,075–87. Oldenburg added the following explanatory title: ... "New Theory about *Light* and Colors: Where *Light* is declared to be not Similar or Homogeneal, but consisting of difform rays, some of which are more refrangible than others: And *Colors* are affirm'd to be not Qualifications* of Light, deriv'd from Refractions of natural Bodies (as is generally believed), but Original and Connate* properties, which in divers rays are divers; Where several Observations and Experiments are alledged to prove the said Theory."

A Letter of Mr. Isaac Newton, *Professor of the Mathematicks in the University of Cambridge, containing his New Theory about* Light *and* Colors: *sent by the Author to the Publisher from Cambridge, Febr. 6. 1671/72, in order to be communicated to the* R. Society.

SIR,

To perform my late promise to you I shall without further ceremony acquaint you that in the beginning of the Year 1666[1] (at which time I applyed my self to the grinding of Optick glasses of other figures than *Spherical*), I procured me a Triangular glass-Prisme, to try therewith the celebrated *Phænomena of Colours*.[2] And in order thereto having darkened my chamber, and made a small hole in my window-shuts* to let in a convenient quantity of the Suns light, I placed my Prisme at his entrance that it might be thereby refracted to the opposite wall. It was at first a very pleasing divertisement to view the vivid and intense colours produced thereby, but after a while applying my self to consider them more circumspectly I became surprised to see them in an *oblong* form, which, according to the received laws of Refraction,[3] I expected should have been *circular*.

They were terminated at the sides with streight lines, but at the ends the decay of light was so gradual that it was difficult to determine justly what was their figure; yet they seemed *semicircular*.

Comparing the length of this coloured *Spectrum* with its breadth, I found it about five times greater; a disproportion so extravagant that it excited me to a more than ordinary curiosity of examining from whence it might proceed. I could scarce think that the various *Thickness* of the glass, or the termination with shadow or darkness, could have any Influence on light to produce such an effect; yet I thought it not amiss first to examine those circumstances, and so tryed what would happen by transmitting light through parts of the glass of divers thicknesses, or through holes in the window of divers bignesses, or by setting the Prisme without, so that the light might pass through it and be refracted before it was terminated by the hole. But I found none of those circumstances material.* The fashion of the colours was in all these cases the same.

Then I suspected whether by any *unevenness* in the glass or other contingent* irregularity these colours might be thus dilated. And to try this I took another Prisme like the former, and so placed it that the light, passing through them both, might be refracted contrary ways, and so by the latter returned into that course from which the former had diverted it. For by this means I thought the *regular* effects of the first Prisme would be destroyed by the second Prisme, but the *irregular* ones more augmented by the multiplicity of refractions. The event was that the light, which by the first Prisme was diffused into an *oblong* form, was by the second reduced into an *orbicular** one with as much regularity as when it did not at all pass through them. So that what ever was the cause of that length 'twas not any contingent irregularity.

I then proceeded to examin more critically what might be effected by the difference of the incidence of Rays coming from divers parts of the Sun; and to that end measured the several lines and angles belonging to the Image. Its distance from the hole or Prisme was 22 foot; its utmost length 13¾ inches; its breadth 2⅝; the diameter of the hole ¼ of an inch; the angle which the Rays, tending towards the middle of the image, made with those lines in which they would have proceeded without refraction, was 44 deg. 56'. And the vertical Angle of the Prisme 63 deg. 12'. Also the Refractions on both sides the

Prisme, that is, of the Incident and Emergent Rays, were as near as I could make them equal, and consequently about 54 deg. 4′.[4] And the Rays fell perpendicularly upon the wall. Now subducting* the diameter of the hole from the length and breadth of the Image there remains 13 Inches the length, and 2⅜ the breadth, comprehended by those Rays which passed through the center of the said hole, and consequently the angle of the hole which that breadth subtended was about 31′, answerable to the Suns Diameter; but the angle which its length subtended was more than five such diameters, namely 2 deg. 49′.

Having made these observations I first computed from them the refractive power of that glass, and found it measured by the *ratio* of the sines 20 to 31. And then by that *ratio* I computed the Refractions of two Rays flowing from opposite parts of the Sun's *discus,** so as to differ 31′ in their obliquity* of Incidence, and found that the emergent Rays should have comprehended an angle of about 31′, as they did before they were incident.

But because this computation was founded on the Hypothesis of the proportionality of the *sines** of Incidence and Refraction, which though by my own Experience[5] I could not imagine to be so erroneous as to make that angle but 31′, which in reality was 2 deg. 49′, yet my curiosity caused me again to take my Prisme. And having placed it at my window as before, I observed that by turning it a little about its *axis* to and fro, so as to vary its obliquity to the light more than an angle of 4 or 5 degrees, the Colours were not thereby sensibly* translated* from their place on the wall, and consequently by that variation of Incidence the quantity of Refraction was not sensibly varied. By this Experiment therefore, as well as by the former computation, it was evident that the difference of the Incidence of Rays, flowing from divers parts of the Sun, could not make them after decussation* diverge at a sensibly* greater angle than that at which they before converged; which being, at most, but about 31 or 32 minutes, there still remained some other cause to be found out from whence it could be 2 degr. 49′.

Then I began to suspect whether the Rays, after their trajection through the Prisme, did not move in curve lines, and according to their more or less curvity tend to divers parts of the wall. And it increased my suspition when I remembred that I had often seen a Tennis ball, struck with an oblique Racket, describe such a curve line.[6] For a circular as well as a progressive motion being communicated to it by that stroak, its parts on that side where the motions conspire must press and beat the contiguous* Air more violently than on the other, and there excite* a reluctancy* and reaction of the Air proportionably greater. And for the same reason, if the Rays of light should possibly be globular bodies, and by their oblique passage out of one medium into another acquire a circulating motion, they ought to feel the greater resistance from the ambient Æther on that side where the motions conspire, and thence be continually bowed to the other.[7] But notwithstanding this plausible ground of suspition, when I came to examine it I could observe no such curvity in them. And besides (which was enough for my purpose) I observed that the difference 'twixt the length of the Image, and diameter of the hole through which the light was transmitted, was proportionable to their distance.

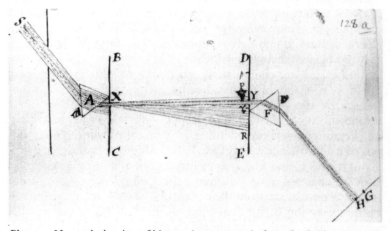

Plate 12   Newton's drawing of his *experimentum crucis*; from Cambridge University Library, MS Add. 4002, fo. 128a

The gradual removal of these suspitions at length led me to the *Experimentum Crucis*,[8] which was this: I took two boards and placed one of them close behind the Prisme at the window, so that the light might pass through a small hole made in it for the purpose and fall on the other board, which I placed at about 12 feet distance, having first made a small hole in it also for some of that Incident light to pass through. Then I placed another Prisme behind this second board so that the light, trajected through both the boards, might pass through that also and be again refracted before it arrived at the wall. This done, I took the first Prisme in my hand and turned it to and fro slowly about its *Axis*, so much as to make the several parts of the Image cast on the second board successively pass through the hole in it, that I might observe to what places on the wall the second Prisme would refract them. And I saw by the variation of those places that the light, tending to that end of the Image towards which the refraction of the first Prisme was made, did in the second Prisme suffer a Refraction considerably greater than the light tending to the other end. And so the true cause of the length of that Image was detected to be no other than that *Light* consists of *Rays differently refrangible**, which, without any respect to a difference in their incidence were, according to their degrees of refrangibility, transmitted towards divers parts of the wall.

When I understood this I left off my aforesaid Glass works, for I saw that the perfection of Telescopes was hitherto limited, not so much for want of glasses truly figured* according to the prescriptions of Optick Authors (which all men have hitherto imagined), as because that Light it self is a *Heterogeneous mixture of differently refrangible Rays.* So that, were a glass so exactly figured as to collect any one sort of rays into one point, it could not collect those also into the same point, which having the same Incidence upon the same Medium are apt to suffer a different refraction. Nay, I wondered that, seeing the difference of refrangibility was so great as I found it, Telescopes should arrive to that perfection they are now at. For, measuring the refractions in one of my Prismes, I found that supposing the common *sine* of Incidence upon one of its planes was 44 parts, the *sine* of refraction of the utmost Rays on the red end of the Colours made out of the glass into the Air would be 68 parts, and the *sine* of refraction of the utmost rays on the other end 69 parts: So that the difference is about a 24th or 25th part of the whole refraction. And consequently the object-glass of any Telescope cannot collect all the rays which come from one point of an object so as to make them convene* at its *focus* in less room than in a circular space whose diameter is the 50th part of the Diameter of its Aperture; which is an irregularity some hundreds of times greater than a circularly figured *Lens*, of so small a section as the Object glasses of long Telescopes are, would cause by the unfitness of its figure, were Light *uniform*.

This made me take *Reflections* into consideration, and finding them regular, so that the Angle of Reflection of all sorts of Rays was equal to their Angle of Incidence, I understood that by their mediation Optick instruments might be brought to any degree of perfection imaginable, provided a *Reflecting* substance could be found which would polish as finely as Glass and *reflect* as much light as glass *transmits*, and the art of communicating to it a *Parabolick* figure* be also attained. But there seemed very great difficulties, and I have almost thought them insuperable when I further considered that every irregularity in a reflecting superficies makes the rays stray 5 or 6 times more out of their due course than the like irregularities in a refracting one: So that a much greater curiosity* would be here requisite than in figuring glasses for Refraction.

Amidst these thoughts I was forced from *Cambridge* by the Intervening Plague, and it was more than two years before I proceeded further.⁹ But then having thought on a tender* way of polishing proper for metall, whereby, as I imagined, the figure also would be corrected to the last,* I began to try what might be effected in this kind, and by degrees so far perfected an Instrument (in the essential parts of it like that I sent to *London*), by which I could discern Jupiters 4 Concomitants,* and shewed them divers times to two others of my acquaintance. I could also discern the Moon-like phase of *Venus*, but not very distinctly, nor without some niceness in disposing the Instrument.

From that time I was interrupted till this last autumn, when I made the other. And as that was sensibly better than the first (especially for Day-Objects), so I doubt not but they will be still brought to a much greater perfection by their endeavours who, as you inform me, are taking care about it at *London*.¹⁰

I have sometimes thought to make a *Microscope* which in like manner should have, instead of an Object-glass, a Reflecting piece of metall. And this I hope they will also take into consideration. For those Instruments seem as capable of improvement as *Telescopes*, and perhaps more, because but one reflective piece of metall is requisite in them, as you may perceive by the annexed diagram, where A B representeth the object metall, C D the eye glass, F their common Focus, and O the other focus of the metall, in which the object is placed.

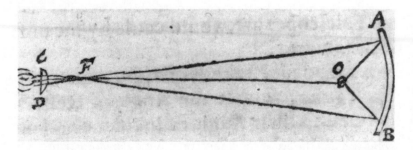

But to return from this digression, I told you that Light is not similar or homogeneal* but consists of *difform** Rays, some of which are more refrangible than others: So that of those which are alike incident on the same medium some shall be more refracted than others, and that not by any virtue of the glass or other external cause, but from a predisposition which every particular Ray hath to suffer a particular degree of Refraction.

I shall now proceed to acquaint you with another more notable difformity in its Rays, wherein the *Origin of Colours* is unfolded.[12] Concerning which I shall lay down the *Doctrine* first and then, for its examination, give you an instance or two of the *Experiments* as a specimen of the rest.

The Doctrine you will find comprehended and illustrated in the following propositions.

1. As the Rays of light differ in degrees of Refrangibility, so they also differ in their disposition to exhibit this or that particular colour. Colours are not *Qualifications* * of Light, derived from Refractions or Reflections of natural Bodies (as 'tis generally believed), but *Original* and *connate* * *properties*, which in divers Rays are divers. Some Rays are disposed to exhibit a red colour and no other, some a yellow and no other, some a green and no other, and so of the rest. Nor are there only Rays proper and particular to the more eminent colours but even to all their intermediate gradations.

2. To the same degree of Refrangibility ever belongs the same colour, and to the same colour ever belongs the same degree of Refrangibility. The *least Refrangible* Rays are all disposed to exhibit a *Red* colour, and contrarily those Rays which are disposed to exhibit a *Red* colour are all the least refrangible. So the *most refrangible* Rays are all disposed to exhibit a deep *Violet Colour*, and contrarily those which are apt to exhibit such a violet colour are all the most Refrangible. And so to all the intermediate colours in a continued series belong intermediate degrees of refrangibility. And this Analogy 'twixt colours and refrangibility is very precise and strict, the Rays always either exactly agreeing in both or proportionally disagreeing in both.

3. The species of colour and degree of Refrangibility proper to any particular sort of Rays is not mutable* by Refraction, nor by Reflection from natural bodies, nor by any other cause that I could yet observe. When any one sort of Rays hath been well parted from those of other kinds it hath afterwards obstinately retained its colour, notwithstanding my utmost endeavours to change it. I have refracted it with Prisms, and reflected it with Bodies which in Day-light were

of other colours; I have intercepted it with the coloured film of Air interceding* two compressed plates of glass; transmitted it through coloured Mediums, and through Mediums irradiated with other sorts of Rays, and diversly terminated it; and yet could never produce any new colour out of it. It would by contracting or dilating become more brisk, or faint, and by the loss of many Rays in some cases very obscure and dark; but I could never see it changed *in specie.**

Yet seeming transmutations of Colours may be made where there is any mixture of divers sorts of Rays. For in such mixtures the component colours appear not, but by their mutual allaying* each other constitute a midling colour. And therefore, if by refraction, or any other of the aforesaid causes, the difform Rays latent in such a mixture be separated, there shall emerge colours different from the colour of the composition. Which colours are not New generated but only made Apparent by being parted, for if they be again intirely mix't and blended together they will again compose that colour which they did before separation. And for the same reason Transmutations made by the convening of divers colours are not real, for when the difform Rays are again severed they will exhibit the very same colours which they did before they entered the composition, as you see, *Blew* and *Yellow* powders, when finely mixed, appear to the naked eye *Green*, and yet the Colours of the Component corpuscles are not thereby really transmuted but only blended. For when viewed with a good Microscope they still appear *Blew* and *Yellow* interspersedly.

5. There are therefore two sorts of Colours. The one original and simple, the other compounded of these. The Original or primary colours are *Red, Yellow, Green, Blew*, and a *Violet-purple*, together with Orange, Indico, and an indefinite variety of Intermediate gradations.

6. The same colours in *Specie* with these Primary ones may be also produced by composition. For a mixture of *Yellow* and *Blew* makes *Green*; of *Red and Yellow* makes *Orange*; of *Orange* and *Yellowish green* makes *yellow*. And in general, if any two Colours be mixed which in the series of those generated by the Prisme are not too far distant one from another, they by their mutual alloy* compound that colour which in the said series appeareth in the mid-way between them. But those which are situated at too great a distance do not so. *Orange* and *Indico* produce not the intermediate Green, nor Scarlet and Green the intermediate yellow.

7. But the most surprising and wonderful composition was that of *Whiteness*. There is no one sort of Rays which alone can exhibit this. 'Tis ever compounded, and to its composition are requisite all the aforesaid primary Colours, mixed in a due proportion. I have often

with Admiration beheld that all the Colours of the Prisme being made to converge, and thereby to be again mixed as they were in the light before it was Incident upon the Prisme, reproduced light intirely and perfectly white, and not at all sensibly differing from a *direct* Light of the Sun, unless when the glasses* I used were not sufficiently clear, for then they would a little incline it to *their* colour.

8. Hence therefore it comes to pass that *Whiteness* is the usual colour of *Light*; for Light is a confused aggregate of Rays indued with all sorts of Colors, as they are promiscuously darted from the various parts of luminous bodies. And of such a confused aggregate, as I said, is generated Whiteness, if there be a due proportion of the Ingredients; but if any one predominate the Light must incline to that colour, as it happens in the Blew flame of Brimstone,* the yellow flame of a Candle, and the various colours of the Fixed stars.

9. These things considered, the *manner* how colours are produced by the Prisme is evident. For, of the Rays constituting the incident light, since those which differ in Colour proportionally differ in Refrangibility, *they* by their unequall refractions must be severed and dispersed into an oblong form in an orderly succession from the least refracted Scarlet to the most refracted Violet. And for the same reason it is that objects, when looked upon through a Prisme, appear coloured. For the difform Rays, by their unequal Refractions are made to diverge towards several parts of the *Retina,* and there express* the Images of things coloured, as in the former case they did the Suns Image upon a wall. And by this inequality of refractions they become not only coloured but also very confused and indistinct.

10. Why the Colours of the *Rainbow* appear in falling drops of Rain is also from hence evident. For those drops which refract the Rays, disposed to appear purple in greatest quantity to the Spectators eye, refract the Rays of other sorts so much less as to make them pass beside it; and such are the drops on the inside of the *Primary* Bow and on the outside of the *Secondary* or Exteriour one. So those drops which refract in greatest plenty the Rays, apt to appear red toward the Spectators eye, refract those of other sorts so much more as to make them pass beside it; and such are the drops on the exteriour part of the *Primary* and interiour part of the *Secondary* Bow.

11. The odd Phænomena of an infusion of *Lignum Nephriticum,*[12] *Leaf gold, Fragments of coloured glass,* and some other transparently coloured bodies, appearing in one position of one colour and of another in another, are on these grounds no longer riddles. For those are substances apt to reflect one sort of light and transmit another, as may be seen in a dark room by illuminating them with similar or

uncompounded light. For then they appear of that colour only with which they are illuminated, but yet in one position more vivid and luminous than in another, accordingly as they are disposed more or less to reflect or transmit the incident colour.

12. From hence also is manifest the reason of an unexpected Experiment, which Mr. *Hook* somewhere in his *Micrography*[14] relates to have made with two wedg-like transparent vessels, fill'd the one with a red, the other with a blew liquor: namely, that though they were severally* transparent enough yet both together became opake, For if one transmitted only red and the other only blew no rays could pass through both.

13. I might add more instances of this nature but I shall conclude with this general one, that the Colours of all natural Bodies have no other origin than this, that they are variously qualified* to reflect one sort of light in greater plenty than another. And this I have experimented in a dark Room by illuminating those bodies with uncompounded light of divers colours. For by that means any body may be made to appear of any colour. They have there no appropriate colour but ever appear of the colour of the light cast upon them, but yet with this difference, that they are most brisk and vivid in the light of their own daylight-colour. *Minium* appeareth there of any colour indifferently with which 'tis illustrated*; but yet most luminous in red, and so *Bise** appeareth indifferently of any colour with which 'tis illustrated, but yet most luminous in blew. And therefore *Minium* reflecteth Rays of any colour, but most copiously those indued with red; and consequently when illustrated with day-light, that is, with all sorts of Rays promiscuously blended, those qualified with red shall abound most in the reflected light and by their prevalence cause it to appear of that colour. And for the same reason *Bise*, reflecting blew most copiously, shall appear blew by the excess of those Rays in its reflected light; and the like of other bodies. And that this is the intire and adequate cause of their colours is manifest, because they have no power to change or alter the colours of any sort of Rays incident apart, but put on all colours indifferently with which they are inlightned.

These things being so, it can be no longer disputed whether there be colours in the dark, nor whether they be the qualities of the objects we see, no nor perhaps whether Light be a Body.[14] For since Colours are the *qualities* of Light,[15] having its Rays for their intire and immediate subject, how can we think those Rays *qualities* also unless one quality may be the subject of and sustain another; which in effect is to call it *Substance*. We should not know Bodies for substances were it not for their sensible qualities, and the Principle of those being now found due to something else, we have as good reason to believe that to be a Substance also.

Besides, whoever thought any quality to be a *heterogeneous* aggregate, such as Light is discovered to be? But to determine more absolutely what Light is, after what manner refracted, and by what modes or actions it produceth in our minds the Phantasms of Colours, is not so easie. And I shall not mingle conjectures with certainties.

Reviewing what I have written, I see the discourse it self will lead to divers Experiments sufficient for its examination: And therefore I shall not trouble you further than to describe one of those, which I have already insinuated.*

In a darkened Room make a hole in the shut of a window, whose diameter may conveniently be about a third part of an inch, to admit a convenient quantity of the Suns light: And there place a clear and colourless Prisme to refract the entring light towards the further part of the Room, which, as I said, will thereby be diffused into an oblong coloured Image. Then place a *Lens* of about three foot radius (suppose a broad Object-glass of a three foot Telescope), at the distance of about four or five foot from thence, through which all those colours may at once be transmitted and made by its Refraction to convene at a further distance of about ten or twelve feet. If at that distance you intercept this light with a sheet of white paper you will see the colours converted into whiteness again by being mingled. But it is requisite that the *Prisme* and *Lens* be placed steddy, and that the paper on which the colours are cast be moved to and fro; for by such motion you will not only find at what distance the whiteness is most perfect, but also see how the colours gradually convene and vanish into whiteness, and afterwards having crossed one another in that place where they compound Whiteness are again dissipated* and severed, and in an inverted order retain the same colours which they had before they entered the composition.* You may also see that if any of the Colours at the *Lens* be intercepted, the Whiteness will be changed into the other colours. And therefore, that the composition of whiteness be perfect, care must be taken that none of the colours fall besides the *Lens*.

In the annexed design of this Experiment, A B C expresseth the Prism set endwise to sight, close by the hole F of the window E G. Its vertical Angle A C B may conveniently be about 60 degrees. *M N* designeth* the *Lens*; Its breadth $2\frac{1}{2}$ or 3 inches. S F one of the streight lines in which difform Rays may be conceived to flow

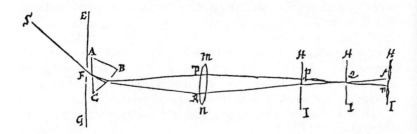

successively from the Sun. F P, and F R two of those Rays unequally refracted, which the *Lens* makes to converge towards Q, and after decussation to diverge again. And H I the paper, at divers distances, on which the colours are projected: which in Q constitute *Whiteness*, but are *Red* and *Yellow* in R, r, and f, and *Blew* and *Purple* in P, p, and π.

If you proceed further to try the impossibility of changing any uncompounded colour (which I have asserted in the third and thirteenth Propositions), 'tis requisite that the Room be made very dark, least any scattering light mixing with the colour disturb and allay it, and render it compound, contrary to the design of the Experiment. 'Tis also requisite that there be a perfecter separation of the Colours than, after the manner above described, can be made by the Refraction of one single Prisme, and how to make such further separations will scarce be difficult to them that consider the discovered laws of Refractions. But if tryal shall be made with colours not throughly separated there must be allowed changes proportionable to the mixture. Thus if compound Yellow light fall upon Blew *Bise* the Bise will not appear perfectly yellow but rather green, because there are in the yellow mixture many rays indued with green, and Green, being less remote from the usual blew colour of Bise than yellow, is the more copiously reflected by it.

In like manner, if any one of the Prismatick colours, suppose Red, be intercepted on design to try the asserted impossibility of reproducing that Colour out of the others which are pretermitted;* 'tis necessary either that the colours be very well parted before the red be intercepted, or that together with the red the neighbouring colours into which any red is secretly dispersed (that is, the yellow, and perhaps green too) be intercepted, or else that allowance be made for the emerging of so much red out of the yellow green, as may possibly have been diffused and scatteringly blended in those colours. And if these things be observed, the new Production of Red or any intercepted colour will be found impossible.

This, I conceive, is enough for an Introduction to Experiments of this kind; which if any of the R. *Society* shall be so curious as to prosecute I should be very glad to be informed with what success: That if any thing seem to be defective or to thwart* this relation, I may have an opportunity of giving further direction about it, or of acknowledging my errors, if I have commited any.[16]

Appendix: Joseph Glanvill's stylistic revisions, 1661 and 1676

1. From *The Vanity of Dogmatizing* (London, 1661). Glanvill is arguing that many things in nature lie beyond human knowledge and its "Mechanical solutions".

And though the fury of that *Apelles*,[1] who threw his Pencil in a desperate rage upon the Picture he had essayed to draw, once casually effected those lively representations which his Art could not describe, yet 'tis not likely that one of a thousand such præcipitancies should be crowned with so unexpected an issue. For though *blind matter* might reach some elegancies in individual effects, yet specifick conformities* can be no *unadvised* productions, but in greatest likelyhood are regulated by the immediate efficiency of some *knowing* agent; which whether it be *seminal* * *Forms*, according to the *Platonical* Principles, or what ever else we please to suppose, the manner of its working is to us *unknown*. Or if these effects are meerly *Mechanical*,* yet to learn the method of such operations may be, and hath indeed been ingeniously attempted; but I think cannot be performed to the satisfaction of severer examination.

That all bodies both *Animal, Vegetable*, and *Inanimate* are form'd out of such particles of matter, which by reason of their figures will not cohære or lie together but in such an order as is necessary to such a specifical formation, and that therein they naturally of themselves concurre and reside, is a pretty conceit, and there are *experiments* that credit it. If, after a decoction* of *hearbs* in a Winter-night, we expose the liquor to the frigid air, we may observe in the morning under a crust of Ice the perfect appearance both in *figure* and *colour* of the *Plants* that were taken from it. But if we break the *aqueous Crystal* those pretty *images* dis-appear and are presently dissolved.

Now these *airy Vegetables* are presumed to have been made by the reliques of these *plantal emissions* * whose avolation* was prevented by the *condensed inclosure*. And therefore playing up and down for a while within their liquid prison they at last settle together in their natural order, and the *Atomes* of each part finding out their proper place, at length rest in their methodical Situation till by breaking the *Ice* they are disturbed, and those counterfeit *compositions* are scatter'd into their first *Indivisibles*. This *Hypothesis* may yet seem to receive further confirmation from the artificial *resurrection* of *Plants* from their *ashes*, which *Chymists* are so well acquainted with; And besides, that *Salt* dissolved upon fixation* returns to its affected* *cubes*, the regular figures of *Minerals*, as the *Hexagonal* of *Crystal*, the *Hemisphærical* of the *Fairy-stone*, the *stellar figure* of the stone *Asteria*, and such like ... (pp. 45–7).

Appendix

2. From *Essays on Several Important Subjects in Philosophy and Religion* (London, 1676), Essay 1: "Against Confidence in Philosophy". Glanvill repeats his argument that God created man, as the first cause, but that we cannot know the "*particular Agent* that forms the Body". The Platonists' theories of a "*Plastick Faculty*" or animating soul explains nothing.

There remains now but one account more, and that is the *Mechanical, viz.* That it is done by *meer Matter* moved after *such* or *such* a manner. Be that so, It will yet be said that *Matter* cannot move *it self*: the question is still of the *Mover*. The Motions are *orderly* and *regular*; Query, *Who guides? Blind Matter* may produce an elegant effect for once by a great Chance; as the Painter accidentally gave the Grace to his Picture by throwing his Pencil in rage and disorder upon it. But then *constant* Uniformities, and Determinations to a *kind* can be *no Results* of *unguided Motions.* There is indeed a *Mechanical Hypothesis* to this purpose, That the Bodies of *Animals* and *Vegitables* are formed out of *such* particles of Matter as by reason of their Figures will not lie together but in the order that is necessary to make *such* a Body, and in *that* they naturally concur and rest; which seems to be confirm'd by the *artificial Resurrection* of *Plants* of which *Chymists* speak, and by the regular Figures of Salts and *Minerals*; the *hexagonal* of *Chrystal*, the *Hemi-spherical* of the *Fairy-Stone*, and divers such like. And there is an experiment mentioned by approved Authors that looks the same way. It is, That after a decoction of Herbs in a frosty Night the shape of the Plants will appear under the Ice in the Morning. Which Images are supposed to be made by the congregated *Effluvia* of the Plants themselves, which loosly wandring up and down in the Water at last settle in their natural place and order, and so make up an appearance of the Herbs from whence they were emitted. This account I confess hath something ingenious in it, But it is no solution of the Doubt.

(pp. 11–12)

Two comments may be made, one on the content, the other on the style of this passage. The "experiment" that Glanvill refers to was described by some alchemical writers, especially Quercetanus (Joseph Duchesne, 1544–1609), and was discussed several times by Robert Boyle. His *Certain Physiological Essays and Other Tracts*[2] (London, 1661) included "Two Essays concerning the Unsuccessfulness of Experiments", the first showing "how experiments may miscarry upon the account of the materials employed in trying them", the second reviewing other circumstances that call in question the truth of experiments (p. 334). One such is the claim made by "those learned writers who affirm that if a lixivium [lye] made of the ashes or fixed salt of a burned plant be frozen there will appear in the ice the idea of the same plant..." (p. 337). But experimental trials to confirm the existence of this Platonic "idea" did not produce "the promised success". Even when freezing was carried out with "a strong decoction of wormwood (from which an idea of the plant may be more probably expected) those to whom I shewed it after it was frozen could discern as little like wormwood in it as myself". True, the ice contained crystal "figures", but these appear in many liquids when frozen. Boyle concludes:

And therefore we cannot but think that the figures that are oftentimes to be met with in the frozen lixivium or decoction of a plant will afford but uncertain proof that the idea of each, or so much as of any determinate plant, displays itself constantly in the frozen liquor. And I much fear that most of those that tell us that they have seen such plants in ice have in that discovery made as well use of their imagination as of their eyes. (*ibid.*)

Boyle performed experiments with freezing over a long period, some appearing in *The Sceptical Chymist*,[3] and collected many of them into one of his typically disorganized, miscellaneous works, *New Experiments and Observations Touching Cold*[4] (London, 1665). In one of the several appendices, "An Account of Freezing made in December and January 1662", Boyle again tests "the famous experiment of *Quercetan*, and affirmed by many other chymists", taking five different vegetables, mixing their distilled waters with their own salts, and putting them into vials. But once freezing was completed, "the effect was that neither of them shewed the least resemblance of the plants from which they were extracted, neither figure, nor shew of roots, stalks, branches, nor leaves (but only a lump or heap of small globuli), much less of flower or seed" (1, 705). Elsewhere he records the freezing of three more "decoctions" without there appearing "in the ice any resemblance of the decocted plants" (1, 724). In the *Experimental History of Cold* proper Boyle confronts more alchemists with his practical tests:

It is confirmed by divers eminent writers, and those modern ones, too, that water impregnated with the saline parts of the plants, and afterwards frozen, will exhibit in the ice the shape of the same plant; and the learned, but I fear too credulous *Gaffarel* tells us that this is no rarity, being daily shown by one *Monsieur de la Clave*. (1, 651)

Having already shown that "this experiment, as it is wont to be delivered, is either untrue or very contingent" (dependent on chance; fortuitous) Boyle merely notes that he repeated the experiment with decoctions of vegetables, "but this ice was by no means so figured as the patrons of the tradition promise". Freezing duly produced ice of various shapes, "little sticks" and "thin parallel plates, exceeding numerous, but (as one of our notes expressly informs us) no way in the shape of trees..." (1, 651). Similarly with the assertion of "the learned [Thomas] Bartholinus" in his treatise *De usu nivis*, that a decoction of cabbage-water, when frozen, "will represent a cabbage, the vegetable spirits being ... concentrated by the cold": Boyle tried the experiment several times, "but the ice did not, either to me or others, appear to have any thing in it like a cabbage, or remarkably differing from other ice" (I, 651–2). As for the teaching of Quercetanus "and other spagyrical writers ... about the seminal virtues surviving in the ashes of burnt plants", Boyle casts doubt on that and other "uncertain traditions of the chymists, whose unsatisfactory way of setting down matters of fact" produces distrust of the experimental method (I, 652–3).

In reproducing the alchemists' claim for their experiments in 1676, then, Glanvill showed himself out of touch with the "mechanical science" that he criticizes in the name of his theological scepticism. As for the style of the two passages, whereas everyone sees the considerable abbreviation in the second (the 250 pages of *The Vanity of Dogmatizing* are reduced to about 60 pages in the *Essay*, making allowance for the differing format), it is not the case, as Stephen Medcalf affirms, that "the removal of metaphor is pervasive, so much so as to convert what we may call a 'symbolic' picture of the world into a 'positive' one".[5] I do not find that much difference in Glanvill's world-picture (as he writes himself, "the truth is, I am not grown so much wiser yet as to have alter'd any thing in the main of those conceptions"), nor is the later version so denuded of metaphor. In the passage reprinted, in 1676 we still find him writing in animistic or anthropomorphic terms of "the congregated *Effluvia* of the Plants themselves ... loosely wandring up and down in the Water" before they "settle in their natural place", and he has certainly not followed Sprat's recommendation of the non-Latinate English of craftsmen. Elsewhere in the Essay we find a constant and fluent use of metaphor and analogy. Men "cannot pry into the hidden things of Nature, nor observe the first Springs and Wheels that set the rest in motion" (p. 2). We speculate about immaterial things "which may be seen in their *effects* and *attributes*, by way of reflection; but if, like Children, we run behind the Glass to look for them, we shall meet nothing there but disappointment" (p. 2). As for the controversies over the soul, Glanvill refuses to "stir in the Waters that have been troubled with so much contention" (p. 3). The way in which the soul directs the animal spirits through the body, given the body's many passages, provokes his wonder that "they should not lose their way in such a wilderness" (p. 4), for "unless we allow it a kind of inward sight of every Vein, Muscle, Artery", and their "several Windings and secret channels", it remains "as inconceivable how it should direct such intricate Motions, as that one that was born blind should manage a Game at Chess, or marshal an Army" (p. 5).

Where Hooke, say, writes extended sequences of metaphor, Glanvill's analogies are brief, used – like Bacon's, often – argumentatively, to clinch a point or generalize an issue. Mental images, Glanvill writes, derive from the brain yet take up remarkably little space there: the image "of an Hemisphere of the Heavens cannot have a Subject larger than the pulp of a Walnut, and how can such petty Impressions make known a Body of so vast a wideness without some kind of *Mathematicks* in the Soul?" (p. 7). Discussing such abstract issues Glanvill resorts always to concrete examples, however incongruously, as when reviewing Kenelm Digby's theory that "things are preserv'd in the Memory by *material Images* that flow from them, which having imping'd on the common sense rebound thence into some vacant Cells of the Brain, where they keep their ranks and postures as they entred, till again they are stirr'd, and then they appear to the Fancy as they were first presented" (p. 8). Rejecting this theory Glanvill develops the implication of the metaphor:

> how is it that when we turn over those Idaea's that are in our memory, to look for any thing we would call to mind, we do not pull all the Images into a disorderly floating, and so make a Chaos of confusion there, where the exactest Order is required? And indeed, according to this account I cannot see but that our Memories would be more confused than our Dreams, and I can as easily conceive how a heap of Ants can be kept to regular and uniform Motions. (pp. 8–9)

Glanvill excels at the dismissive analogy: "these Controversies, like some Rivers, the further they run the more they are hid" (pp. 9–10). When we leave the evidence of the senses "and retire to the *abstracted notions* of our minds, we build Castles in the Air, and form *Chymerical Worlds*, that have *nothing real* in them" (p. 17; see Bacon, *Works*, II, 337, IV, 32–3).

The inspiration for some of Glanvill's metaphors, as for so many seventeenth-century writers, was Francis Bacon. Glanvill, like Bacon, attacks slavish reliance on the ancients, for "Our Discoveries, like water, will not rise higher than their Fountains" (p. 25). Bacon had used this metaphor many times: "For as water will not ascend higher than the level of the first spring-head from whence it descendeth, so knowledge derived from Aristotle, and exempted from liberty of examination, will not rise again higher than the knowledge of Aristotle" (III, 290, 227; IV, 16). Indeed Glanvill once, unusually, declares his indebtedness: "For as the forementioned great Man, the

Lord *Bacon* hath observ'd, *Time*, as a *River*, brings down to us what is more *light* and *superficial*, while the Things that are more *solid* and *substantial* are sunk and lost" (p. 28; see Bacon, *Works*, IV, 76). Baconian, too, is the judgment that if we study the history of science from its beginning "we shall find that though it hath often changed its Channel, removing from one Nation to another, yet it hath been neither much *improved*, nor *altred*, but as Rivers are in passing through different Countries, *viz*. in *Name*, and *Method*" (p. 27).

Glanvill may have pronounced himself in 1665 dissatisfied with the "curiosity of *fine Metaphors*", but in 1676 he was using metaphors (whether he would now classify them as "fine") as frequently and fluently as any of his contemporaries. One passage that puts this question beyond dispute is his climactic attack on dogmatism:

Dogmatizing shews *Poverty* and *narrowness* of Spirit. There is no greater *Vassallage* than that of being enslaved to Opinions. The *Dogmatist* is pent up in his Prison, and sees no Light but what comes in at those Grates. He hath no *liberty* of *Thoughts*, no *prospect* of various *Objects*; while the *considerate* and *modest* Inquirer hath a *large* Sphere of Motion, and the satisfaction of more *open* Light. He sees *far*, and injoys the pleasure of surveying the *divers* Images of the Mind. But the *Opiniator* hath a *poor shrivel'd* Soul, that will but just hold his little Set of Thoughts. His Appetite after Knowledge is satisfied with his few *Mushromes*, and neither knows nor thinks of any thing beyond his Cottage and his Rags. (p. 32)

Glanvill may have felt "fettered and tied" in his new and more rigorous manner of writing, ruthlessly abridging an essay fifteen years old, and hence lacking "that ease, freedom and fullness" that he might have enjoyed had he been starting afresh, but we must guard against taking such comments as absolute and considered stylistic judgments. They are often conventional palinodes, making the right sound but never put into practice. Above all we must not take them as accounts of style without bothering to examine the style for ourselves. Whatever "plain style" meant in this period, for many writers it still included metaphor and imaginative appeal.

Glossary

absconded: hidden

accident: unfavourable symptom; non-essential quality or attribute

accumulation: piling up

additament: addition

adepti: esoteric initiates

admirable: to be wondered at

advantageous: furthering progress; beneficial

adventitious: casual, accidental

affection: state

agat: precious stone, one of the semi-pellucid variegated chalcedonies

airs: climates

allaying: modifying

alloy: combination, tempering

alumen plumosum: ferroso-aluminic sulphate

ambient: surrounding

Amianthus: a mineral, a variety of asbestos, composed of long pearly white fibres

angulization: becoming angulate, with angles or corners

animal-spirits: animating principle; blood

animated: endowed with life

animation: state of being alive as a human being

annexed: added

aphorismes: scientific laws

apish: imitative

application: remedy, treatment

appropriated: suitable; special

Aqua fortis: nitric acid, or other powerful dissolvent

arcana: secrets

arches: curvatures, contours

architectonick: directing, formative

arefaction: drying up

armed: Lat. '*in armato magnete*': fitted with an armature, a piece of soft iron placed in contact with the poles which preserves and increases the magnetic power

artificer: inventor

artist: practitioner (especially of the 'art' of alchemy)

asperity: unevenness of surface

author: creator, maker

automata: living beings, viewed materially

avolation: evaporation

axiome: intermediate generalization of scientific laws

bank: store

basilisks: artillery

benumming: benumbing, paralyzing

bise: blue bice (a carbonate of copper)

black ground: background (textile)

blas: Van Helmont's term for a supposed flatus or influence of the stars, affecting the weather

blazing: proclaiming

blowne furnaces: blast furnaces

bottom: clew on which to wind thread; keel of a boat

bow-dye: a scarlet dye

box: box-tree, a small evergreen

brim-stone: sulphur

broath: decoction, thin soup

brooking: enduring

bur-dock: coarse weedy plant

calcined: reduced by fire to a powder

calcining: burning to ashes

caldera: cauldron-like cavity on the summit of an extinct volcano

candying: encrustment

canting: secret jargon of a sect; using current stock-phrases

canvass: to scrutinize, examine

Cap-a-pe: from head to foot

captious: fault-finding

caput mortuum: worthless residue

Glossary

card: iron instrument with teeth, used to set in order the fibres of wool

case-wing'd: wings having an outer protective covering

casualty: chance occurrence

catamount: leopard or panther

catholic: general

chafer: cockchafer, winged beetle

chaps: cracks, open fissures

characteristick: indication

charge: cost, expense

charr'd: burnt; reduced to charcoal

chop: jaw

chylus: chyle, "the white milky fluid formed by the action of the pancreatic juice and the bile on the chyme" (OED)

circle: dial

circuites: tours

clear: to confirm, convince

clea's: claws

clew: ball or thread of yarn; clue

clitch: to close, contract

cloth'd: embodied, invested

coagulation: thickening

coalition: combination, fusion

cobb: the head of a red herring

coherence: cohesion

collect: to gather, focus

collection: discussion

colours: rhetorical modes or figures

comminuted: reduced to minute particles

commixture: mixture

common place: category or head of argument

composition: compounds of metals or other elements; combination; union

conceipt: conceive

concomitant: satellite, companion

concrete(s): solid(s)

concreted: solidified

condensation: compression

confiture-house: factory for making sweet-meats

conformity: correspondence, congruity

confounding: confusing

confusion: chaos

congelation: freezing

congeries: mass, pile

conical sections: glass cutting

connate: innate, inborn

conspicuous: clearly visible, open

constitution: composition

contexed: woven together

contiguous: adjoining

contingency: uncertainty

contingent: accidental

convene: come together, converge

conversation: society

convulsive disorders: ailment or illness accompanied by a convulsion

cracker: fire-work

critical: exact, punctual

crooks: curved extremities

curiazier: soldier wearing breast-armour

curiosity: care, carefulness; accuracy

curious: elaborate; careful

daucus: carrot

debilitated: weakened

declination: decay, decline

decoction: extract

decrement: decrese, diminution

decussation: crossing of lines

degrees: steps

delineated: illustrated

demonstrate: exhibit

depredatour: summarizer, maker of abstracts

deserted: relinquished

designeth: designates

diachylon: sealing compound of organic and inorganic substances, solidifying when heated

difform: differing in shape

dioptricks: construction of lenses (the microscope)

disbanded: separated

disclaimed: repudiated

discover: reveal

discus: disk

disgust: dislike

dispatch: speed; dismissal, settlement; to make haste; get rid of, kill

display: spread out

disquisition: investigation

dissipated: split up, divided, disintegrated

distemper: disorder

divination: prognostication or diagnosis; warning

domestick receipts: private techniques

dowry-men: endowment men, sources of wealth

dress'd: prepared, cooked

drossie: dreggy (earth being regarded as opposed to air)

ducket: ducat
ductility: capability of being extended by beating
ductus: movement, stroke

ebullition: sudden outburst or outgrowth
Echini: of the sea-urchin; the skin being covered with spines
effable: able to be expressed in words
efficient: agent
effluviums: the (real or supposed) outflow of material particles too subtle to be perceived by the senses
electuaries: a medicine in which a dry ingredient is mixed with honey, jam, or syrup
elixir: the *elixir vitae* or philosopher's stone
emission: radiation, exhalation
empiric: one who only observes with the senses
empirically: based on the results of observation and experiment
engine: machine
engine-house: engineering-shop
esteemed: estimated
exactor: one who insists on, or exacts
exception: objection
excite: arouse, encourage
exhalation: evaporation
expansum: expanse (firmament)
experience: experiment(s)
express: represent
exquisitely: carefully, minutely
exsanguineous: bloodless
exsuction: action of sucking out
extension: extent, size
extenuate: made thin
extravagance: divergence, digression

fabric: frame, structure
fairly: clearly
faithful: accurate
fallace: refutation, exposure
fermentation, long: fermented for a long time
figure: shape; give figure to
fire, in the: using a chemical furnace
fixation: becoming solid
flat: cross-wise
flitter'd: fragmented
foraminulous: full of holes
founder: fall

frame: loom; timbers constituting a ship's body
Free-stone: fine-grained sandstone or limestone
fret: to wear away
fretting: friction, irritation
friability: the quality of being easily crumbled or reduced to powder
frow: brittle
fuliginous: sooty

gas: Van Helmont's term for a supposed occult principle in all bodies, an ultra-rarefied condition of water
gelly: jelly
generation: reproduction
Ginny: Guinea
glass: microscope, lens
glaz'd: glass
gluten: sticky substance
grater: instrument with a rough indented surface used for grating or rasping
gravers: engravers
guilt: gilded

Halcyon: quiet, peaceful
harangue: vehement address, tirade
hard word: technical term
harsh: hard
hartshorn: a solution of ammonia in water
heap: collection
heat: quarrel
historical: factual
history: natural history, not necessarily chronological
homogeneal: homogeneous
humour: nature, temperament; fluid
hydrargyral: mercurial

illiberal: not intellectual; involving manual activities
illustrate: clarify
illustrated: illuminated
impertinent: irrelevant
impetuosity: violent energy
imposers: taskmasters
impression: mark, indentation
incorporate: combine
incumbent: exerting downward pressure
indenture: notch, incision
induration: hardening

Glossary

industry: inventiveness, endeavour
inoculating: budding of trees
inoculatour: one who buds plants
inquisitive: those doing research
insensible: cannot be felt
inservient: serving
insinuate: suggest, introduce
insipid: without taste
in specie: in kind, i.e. in
   itself
inspection: examination
institution: custom, usage
instrument: means, cause
interceding: coming between
interest: involvement
interrogatories, upon: by interrogation
interstitia: intervening spaces
Irish-stitch: embroidery
irruption: bursting or breaking in
issue: termination, conclusion

jealous: vigilant, watchful
jemmar'd: hinged

knob: rounded protuberance

laborant: laboratory-assistant
lanthorne: lantern
Lapidescent: in the process of becoming
   stone, from petrifying waters
largesse: gift, present
last, to the: to the utmost
lattice: structure made up of bars,
   intersecting horizontally or
   diagonally
leavening: fermentation
Lignum Fossile: fossilized wood
limber: flexible
liquor: fluid
load-stone: magnet
luciferous: light-bringing (one of
   Bacon's two main categories for
   experiments)
luted: closed, sealed
lymphiducts: lymph vessels and glands

marchasite: pyrite
material: pertinent, relevant
mawes: inward parts
mechanical: produced by the interplay
   of matter and motion
mechanisme: design, structure
meer: nothing less than
menstruum: solvent; liquid agent

mesenterical: belonging to the intestinal
   canal
meum et tuum: mine and thine; the
   rights of property
minium: red lead
mixt: compound
modesty: moderation, freedom from
   excess
module: model
moity: half
morose: fastidious, fussy
mortified: made tender
motion: working
motion of return: pendulum
multiply: intensify
muscle: mussel
Muscovy-glasse: common mica; silicate
mutable: changeable, affected
mystery-men: those who study crafts

namesetter: one who classifies and
   orders
naturalist: natural philosopher, scientist
nepetides: neap tides
nice: needing great precision
nicety: minuteness, delicacy
non-ultra: "no beyond", the slogan
   erected on the pillars of Hercules set
   in the Straits of Gibraltar, marking
   the limits of the known world

obliquity: deviation, angle
obnoxious: exposed, open to censure,
   reprehensible
obtus'd: blunt, not pointed
opened: dissected
ophthalmia: inflammation of the eye
orb: orbit
orbicular: spherically shaped
orbiculation: acquiring a circular form
otocousticons: ear-trumpets
outside: deceiving external
outvye: exceed

palm: flat widened part at the end of an
   arm or leg
palysado'd: fenced
parallelepiped, four-sided: prism whose
   base is a parallelogram
parallelepiped, rhomboidical: prism
   whose base is a rhomboid
parcel: part, portion
particle: small part
patten: sole (in humans, forming the

base of an overshoe)
peculiar: individual, specific
pellucid: translucent, transparent
pension: payment
pervious: permeable
philosophers's stone: "reputed solid
 substance supposed by alchemists to
 have the property of changing other
 metals into gold or silver, and of
 prolonging life indefinitely". (OED)
philosophical: scientific
phisike: both surgery and medicine
phlegm: mucus; viscous secretion
physic: medicine
physicians: scientists and doctors
pike: peak
pile: group, collection of things
pinked: punched
plain: plane
plantanimation: state of being a plant
plastick vertue: formative power
polish: to refine
pompion: pumpkin
posture: position
Praecognita: something necessary to be
 known beforehand as a basis of
 investigation
praesage: predict
predicament: logical category
prejudice: damage
premise: introduce
premis'd: stated in the premises
prescience: foresight, prior knowledge
prescription: use, custom
pretergenerations: preternatural
 generations, monstrous births
pretermitted: excluded, passed over
prevent: anticipate
primitive: original
primogeneal: first generated; primary
principles: fundamental qualities
pristine: original
privation: absence of quality
probation: proof
proprieties: properties
prosecute: treat in more detail
prospect: description, aspect
publick treasure: treasury
punctual: precise, exact
punctually: in great detail; carefully

quadrare: square with, conform to
qualification: alteration, modification
qualifie: attemper, adapt

qualified: endowed with qualities;
 limited in some respect; imperfect;
 controlled, regulated
qualities: attributes, properties
quality, of: noble-born
quartans: fever, characterized by a
 cramp every 4th day
quill: small tube

ramification: subdivision
ranger: one who classifies and orders
rarefy: lessen the density
rate, at the – of: appropriate to
ratiotination: reasoning
reckoning: estimate of a ship's position
 from the distance run by the log and
 the courses steered by the compass,
 but without astronomical observations
recon'd: reckoned on, counted on
reconditory: repository
rectified spirit: alcohol concentrated by
 distillation
reduce: take back; arrange, order
reduc'd: set down, arranged; brought
refrangible: capable of being refracted
register: record accurately;
 distinguish
relation: report, narrative
reluctancy: resistance
remedies: cures
remoras: obstacles
rent: tear, split
repleated: filled, provided
represent: argue
rest on: depend, rely on
restagnant: overflowing
restitution: tendency to return to a
 previous position by virtue of
 elasticity
rise: basis
rode: path, way
rosin: resin
roving: random
rugosities: roughness, inequalities
running: fluid

sable: black
salgem: rock-salt
sallet: salad
scantlings: prescribed size
scaple: scapple (make a plane surface)
schematisme: inner structure
scheme: illustration

Glossary

scholar: one who studies in the "schools" at a university, and knows Latin; academic

scholemen, schoolmen: scholastics, followers of Aristotle; medieval philosophers

scholes, schools: universities (where Aristotelian science still prevailed)

scientifical: designed for the furthering of knowledge

sea-mark: navigation warning

secur'd: acquired, ensured

seed: cause; formative principle

seminal: influencing others in a new way

sensible: perceptible, evident; aware

sensibly: perceptibly, visibly

separations: methods of separating the elements of a substance

sequestrable: capable of being separated

serv'd: treated

severally: separately

sheath'd and crustaceous: with a hard, close-fitting shell or crust

shot: formed

simple: medicinal herb

sines of Incidence: Snel's law

slide: quarter-tone and lesser musical interval

sliffe: mica

snuff: wick

sodder: to solder

sodered: soldered

sons of art: initiates

sophisticated: adulterated, impure

sorites: series

sower: sour

spagyrist: alchemist

spangle: small round piece of glittering metal

spar: crystalline mineral

specious: plausible, convincing

speculative: theoretical, unfounded in reality

spheres of pasteboard: globes made of cardboard

sphincter: contractile muscle

spirit: distilled extract

spittle: saliva

sport: abnormality, exception

springy: elastic

State: canopied throne

stercorary: living or feeding on dung

stiriae: icicles, small drops

stopple: stopper, plug

straitned: confined

stress: force

subducting: subtracting

subjacent: underlying

subscrib'd: submitted, conformed

subtility: delicacy

subversion: overturning

supply: to compensate for; making good

surmount: exceed, predominate

suture: junction

swelling: amplified, exaggerated; excess, puffing up

swimming-girdle: lifebelt

symbolize: resemble, agree with

talk: talc, mica

tallons: claws of a bird

tartar: potash

teasel: plant of the genus *Dipsacus*, comprising herbs with prickly leaves and flower-heads

tell: count

temper: proportion, balance

tenacious: cohesive

tender: cautious; delicate, exact

tenent: doctrine, dogma

tenter: hook

terminus a quo: starting-point

terraqueous: terrestrial

thorow: through

thwart: contradict, be at variance with; obstruct

tincted upon: tinctured with

tippett: cape trimmed with fur

Topics: logical work itemizing the categories from which arguments may be taken

torture: analyse

touchstone: black or dark-coloured variety of quartz or jasper

tract: course

transcendental: metaphysical

translated: moved

transpeciate: transform

trellice: with bars intersecting horizontally or diagonally

tria prima: three prime substances

trod on: frequently met

truly figured: accurately made

trunk: speaking tube

Tunica Cornea: corneal membrane

Tunica Uvea: curved membrane

turgent: swollen
turn-pikes: spiked barriers

unadorned: unelaborated
uneasy: difficult
unsevered: united
untwist: resolve
us'd to: made a practice of

variation: irregularity
vegetative: endowed with the power of
    growth
vertue: property; power, efficacy
vibrissant: vibrating
virtuoso: scientist
vitrificated: turned into glass
vitrioll: the hydrous sulphate of a
    metal
volatilize: render insubstantial
vorage: chasm, gulf
vortexes: in Descartes' theory of the
    universe, "supposed rotatory

movements of cosmic matter around
    a centre or axis, regarded as
    accounting for the origin or
    phenomena of the terrestrial or other
    systems" (*OED*)
vulcan: volcano
Vulcan, disciples of: alchemists, users of
    the furnace
vulgar: common; vernacular

warm: excited
warned forth: sent out
wasting: decay
waved: undulating
whether: whither
wilde-fires: inflammable material, like
    bitumen or napalm
wind-gun: gun using compressed air
window-shuts: shutters
work, the great: the production of the
    philosopher's stone
wreath'd: plaited, coiled
wrought: impressed, affected

Notes

PREFACE

1 Charles Webster, *Samuel Hartlib and the Advancement of Learning* (Cambridge, 1970); *The Great Instauration. Science, Medicine and Reform 1626–1680* (London, 1975).

2 I have argued this thesis more fully in "The Royal Society and English Prose Style: A Reassessment", in *Rhetoric and the Pursuit of Truth* (Los Angeles: The Clark Library, 1985), pp. 1–76.

3 See, e.g., P. B. Medawar, "Is the Scientific Paper a Fraud?", *The Listener*, 12 Sept. 1963 (Vol. 70), pp. 377–8, and *The Art of the Soluble* (New York, 1969), p. 7.

INTRODUCTION

1 See *The Works of Francis Bacon*, ed. James Spedding, R. L. Ellis, and D. D. Heath, 14 vols. (London, 1857–1874), vol. VIII, pp. 123–43, 332–42, and 376–86, for the early masques. All future references to this edition will be included in the text in the form (III, 237).

2 On Bacon as a scientist see Paolo Rossi, *Francis Bacon. From Magic to Science*, tr. S. Rabinovitch (London, 1968); B. Farrington, ed. and tr., *The Philosophy of Francis Bacon* (Liverpool, 1964); and Mary Hesse, "Francis Bacon's Philosophy of Science", repr. in Brian Vickers (ed.), *Essential Articles for the Study of Francis Bacon* (Hamden, Conn., 1968; London, 1972), pp. 114–39.

3 See Brian Vickers, "Bacon's so-called 'Utilitarianism': sources and influence", in Marta Fattori (ed.), *Francis Bacon. Terminologia e Fortuna nel XVII Secolo* (Rome, 1984), pp. 281–313.

4 On Bacon as a writer see Brian Vickers, *Francis Bacon and Renaissance Prose* (Cambridge, 1968); Lisa Jardine, *Francis Bacon. Discovery and the Art of Discourse* (Cambridge, 1974).

5 A. R. Hall, *The Revolution in Science 1500–1700* (London, 1983), p. 194. Elsewhere Hall presents evidence of Bacon's influence on major projects in England, France, and Germany: pp. 212, 217, 222, 225–6, 228, 230.

6 On Bacon's influence see R. F. Jones, *Ancients and Moderns. A Study of the Rise of the Scientific Movement in Seventeenth-Century England* (St. Louis, Mo., 1936; rev. ed. 1961), and Charles Webster, *The Great Instauration*. Further documentary evidence can be obtained from R. W. Gibson, *Francis Bacon. A Bibliography of his Works and of Baconiana to the year 1750* (Oxford, 1950), and *Supplement* (Oxford, 1959).

7 *The Advice of W.P. to Mr. Samuel Hartlib, for the Advancement of some Particular Parts of Learning* (London, 1648), pp. 2, 13; cited by R. F. Jones, pp. 89, 293.

8 In *The Posthumous Works of Robert Hooke*, ed. Richard Waller (London, 1705), pp. 1–70.

9 *Experimental Philosophy* (London, 1664), p. 82.

10 Cited by R. F. Jones, *Ancients and Moderns*, p. 221.
11 Boyle, *Works*, ed. T. Birch, 2nd ed., 6 vols. (London, 1772), I, 305.
12 *Ibid.*, II, 243; II, 57; V, 514. On Boyle's debt to Bacon see Marie Boas, "Boyle as a Theoretical Scientist", *Isis*, 41 (1950), pp. 261–4; and Baden Teague, *The Origins of Robert Boyle's Philosophy*, Cambridge University Ph.D. thesis, 1972, no. 8218.
13 Boyle, *Works*, II, 1–246; III, 392–414.
14 *Ibid.*, I, xxxiv, xxxvi, xl, xlvii.
15 *Ibid.*, VI, 86, 88.
16 R. F. Jones, *Ancients and Moderns*, pp. 171–6, and G. Keynes, *A Bibliography of Dr. Robert Hooke* (Oxford, 1960), pp.2–4.
17 See Rossi, *Francis Bacon. From Magic to Science*, in note 2, chapter 1.
18 Cited by Michael Hunter, *Science and Society in Restoration England* (Cambridge, 1981), p. 37. This is the most comprehensive recent study of the Royal Society and science in this period, although the very fullness of coverage and the author's tendency to side with all parties sometimes obscure the issue involved.
19 Sprat, *History*, pp. 153 ff.; Glanvill, *Plus Ultra* (London, 1668); Hooke, "A Vindication of the Royal Society", included in one of his Cutlerian lectures: *Posthumous Works*, ed. R. Waller, pp. 329–45; Wallis, *A Defence of the Royal Society* (London, 1678).
20 See Hunter, pp. 136–61, for a discussion which documents the criticism fully, but fails to clarify the position. The two most prominent critics, Meric Casaubon (1599–1670) and Henry Stubbe (1632–1676), accused the Royal Society of being philistine, responsible for the decay of learning (*ibid.*, pp. 151–2, 156). Yet, as Hunter also notes (without making the connection), there is no evidence of any decay of learning, indeed the very subjects that Casaubon valued flourished during this period (pp. 156–7). Some traditionalists at Oxford and Cambridge clung to Aristotelianism (pp. 138, 149), and feared that the Royal Society was trying to usurp the universities' privileges and powers (pp. 144–6). However, Hunter also gives ample evidence of the influence of the new sciences at both universities (pp. 141–3). The real issues, it seems to me, are, first, the general dispute between the arts and the emergent sciences over which of them were truly "useful" to society, the old tradition of beneficent action for the public good in the *vita activa* (see Vickers, cited in note 3 above). Traditionally, philosophy, history, and rhetoric had occupied this privileged role, from the time of Petrarch to Obadiah Walker in 1673 (cited by Hunter, p. 160); but in the writings of Bacon, Hooke, Boyle, Sprat, and many more, the new sciences had claimed that they were more useful (e.g. Hooke, cited by Hunter, p. 146). Hunter has all the evidence for this interpretation of the controversy, but fails to see its relevance. Stubbe felt that if Sprat's *"History* take place, the whole education of this land and all religion is subverted" (p. 148). The extreme, at times hysterical tone in which Stubbe formulated his objections to the Royal Society tells us more about his personality than about the actual state of affairs. Stubbe invoked Aristotle and Galen against the new sciences (p. 150), as did Casaubon, a more sober, indeed pious theologian and philologist, who described Aristotle as "'that inestimable magazin of human learning', believing 'that there is scarce any art, or faculty, wherein we do not come short of the Ancients'" (p. 149), as did others quoted by Hunter. (It is hard to see how he can then claim that the polemics of Stubbe and Casaubon had nothing to do with the wider dispute between ancients and moderns: p. 159.)
   The other issue was the fear that Stubbe and Casaubon expressed about the Royal Society – or indeed, one feels, any new form of knowledge – as being a threat to the Church of England. Despite Sprat's elaborate attempt to ingratiate that

scientific institution with the church, it was not enough for his critics. Casaubon distrusted any learning other than that which would maintain "the creditt and authoritie of the Scriptures themselves", settle controversies in religion and philological disputes about the authenticity of scriptural texts, and he dismissed science as of no use to this activity (Hunter, pp. 155–6). Stubbe thought that the universities should teach their students only "so much *Philosophy* as is necessary for the explaining and defending of our Religion against *Atheists, Papists,* and Socinians" (p. 153). The Royal Society, Stubbe claimed, was not only opposed to this aim but was actually subverting learning, religion, and even public morals. As he put it,

> Ignorance is epidemical, and insensibly diffuses itself through the gentry and all professions, and common debauchery, atheism and popery have grown more than usually through our virtuosi. (cited by Hunter, p. 152)

Where is the evidence for this? one might expect a historian to ask. Or where is Stubbe's evidence for his claim that the experiments conducted by the Royal Society were "*trivial, defective,* and false", and its scientists "unoriginal, inept and dishonest" (p. 150)? Hunter does not ask these questions, indeed he explicitly says that Stubbe and Casaubon "were right" in attacking the "iconoclasm" of the new science (pp. 151, 152), and accusing them of "philistinism" (p. 158), even though he himself rebuts the latter claim, again without connecting the two parts of his chapter (p. 158). To me it seems that this controversy is another version of the "Disputes of the Liberal Arts" so frequent in the Middle Ages and the Renaissance; but that in the hands of Stubbe it reached the level of a Punch and Judy show. Hunter summarizes Stubbe's fear of "the new science as a Catholic plot to draw Protestants away from the defence of their church so that '*a Popish implicit faith*' could be introduced" (p. 155), but the appropriate word for these fears is the one he uses later to describe seventeenth-century anxieties about "atheism": "paranoia about this threat was as pervasive as the hate of Roman Catholicism which was so powerful in politics" (p. 162). Fears are indeed real to the paranoid, but must history be re-written from their viewpoint? Hunter takes Stubbe's side sympathetically, while James R. Jacob, in *Henry Stubbe, radical Protestantism and the early Enlightenment* (Cambridge, 1983), turns him into an intellectual hero, a remarkable attempt to rewrite history.

I am far from thinking the new science to have been epoch-making by the 1670s, yet it had achieved some things of importance, and many of these criticisms seem to me unfounded, as well as unjustifiably abusive.

21 See R. S. Westfall, *Science and Religion in Seventeenth-Century England* (New Haven, Conn., 1958); R. Hooykaas, *Religion and the Rise of Modern Science* (Edinburgh, 1972).

22 See Marie Boas, "The Establishment of the Mechanical Philosophy", *Osiris*, 10 (1952), pp. 412–541. R. H. Kargon, *Atomism in England From Hariot to Newton* (Oxford, 1966); and Norma E. Emerton, *The Scientific Reinterpretation of Form* (Ithaca, NY, 1984).

23 Hooke, *Posthumous Works*, pp. 171–2.

24 See Brian Vickers, "Analogy versus identity: the rejection of occult symbolism, 1580–1680", in Vickers (ed.), *Occult and Scientific Mentalities in the Renaissance* (Cambridge, 1984), pp. 95–163.

25 Quoted by George Williamson, *Senecan Amble. A Study in Prose Form from Bacon to Collier* (London, 1951), p. 178.

26 Cited by R. F. Jones, *The Seventeenth Century* (Stanford, Ca., 1951), p. 84.

27 Boyle, *Works*, I, 300 ff.

28 Hooke, *Posthumous Works*, pp. 9–11.
29 *Ibid.*, pp. 63–4.
30 *Ibid.*, p. 72.
31 *Ibid.*, pp. 171–2.
32 See A. C. Howell, "*Res et Verba*: Words and Things", *ELH*, 13 (1946), pp. 131–42.
33 *Institutes of Oratory*, VIII, proem, 20–21; tr. H. E. Butler, 4 vols., Loeb Classical Library (London, 1922), III, 189. Donne, *Sermons*, ed. E. Simpson and G. R. Potter (Berkeley and Los Angeles, Ca., 1953–1962), X, 112.
34 Howell, cited in note 32. See also G. A. Padley, *Grammatical Theory in Western Europe, 1500–1700: The Latin Tradition* (Cambridge, 1976), chap. 3, "The Seventeenth Century: Words versus Things", although this account is over-simplified.
35 See Vickers, cited in note 24, pp. 103–5, 109–15.
36 See R. F. Jones, *The Seventeenth Century*, pp. 83–4, 102 note, 123 note.
37 *Poems of Abraham Cowley*, ed. A. R. Waller (Cambridge, 1905), pp. 449–50.
38 See R. F. Jones, essays collected in *The Seventeenth Century*, and Charles Webster, *The Great Instauration*.
39 See Brian Vickers, "The Royal Society and English Prose Style: A Reassessment", Clark Library Seminar (March 1980), now in *Rhetoric and the Pursuit of Truth: Language Change in the Seventeenth and Eighteenth Centuries* (Los Angeles, 1985).
40 Boyle, *Works*, I, xxi; *ibid.*, pp. xxxvii and xlvi for Boyle's references to rhetoric in his letters to Hartlib.
41 *Ibid.*, II, 247–322.
42 *Ibid.*, e.g., II, 101, 122; VI, 142, 168, 213, 293.
43 Hooke, *Posthumous Works*, p. 165.
44 *Ibid.*, p. 131; also p. 141.
45 Boyle, *Works*, e.g., II, 39, 59, 63. On the rhetorical tradition of metaphor as providing both illumination and persuasive force see Vickers, *Francis Bacon and Renaissance Prose*, pp. 141–53.
46 See R. F. Jones, *The Seventeenth Century*, pp. 89–97; Jackson I. Cope, *Joseph Glanvill, Anglican Apologist* (St. Louis, Mo., 1956), pp. 152–9; Stephen Medcalf (ed.), *The Vanity of Dogmatizing: the Three Versions* (Hove, 1970), pp. xxi–xxii, xxxiv–xxxix.
47 On universal language schemes see, e.g., Paul Cornelius, *Languages in Seventeenth- and Early Eighteenth-Century Imaginary Voyages* (Geneva, 1965); James Knowlson, *Universal Language Schemes in England and France, 1600–1800* (Toronto, 1975); and Vivian Salmon, *The Study of Language in Seventeenth-Century England* (Amsterdam, 1979).
48 See Mary Slaughter, *Universal Languages and Scientific Taxonomy in the Seventeenth Century* (Cambridge, 1982).
49 See Charles Raven, *John Ray Naturalist* (Cambridge, 1942), pp. 181–3, 343.
50 R. F. Jones, *The Triumph of the English Language* (Stanford, Ca., 1953); Thomas Finckenstaedt, Ernst Leisi, and Dieter Wolff (eds.), *A Chronological English Dictionary* (Heidelberg, 1970), which lists year by year new words as recorded by *OED*.
51 Browne, *Works*, ed. C. Sayle, 3 vols. (London, 1904–7), I, 117.
52 Charleton, *Deliramenta Catarrhi* (London, 1650), Sig. A₃ʳ; Henry Guerlac, "Can there be colors in the dark? Physical color theory before Newton", *Journal of the History of Ideas*, 47 (1986), pp. 3–20, at p. 16.
53 Evelyn, letter of 20 June 1685 to Sir Peter Wyche, in J. E. Spingarn (ed.), *Critical Essays of the Seventeenth Century*, 3 vols. (Oxford, 1908–9), II, 310–13. The term *logodaedali*, 'word-artificers', is taken from Plato, *Phaedrus* 266e, or Quintilian, *Institutio Oratoriae*, 3.1.11.

54 Digby, *Two Treatises* (1641), cited by R. F. Jones, "The Rhetoric of Science in England of the Mid-Seventeenth Century", in C. Camden (ed.), *Restoration and Eighteenth-Century Literature* (Chicago, 1963), p. 19, n. 25.

55 Boyle, *The Sceptical Chymist*, ed. E. A. Moelwyn-Hughes, Everyman Library (London, 1967), p. 167.

56 See F. R. Johnson, "Latin versus English: The Sixteenth-Century Debate over Scientific Terminology", *Studies in Philology*, 41 (1944), pp. 109–35; and S. V. Larkey, "Scientific Glossaries in Sixteenth-Century English Books", *Bulletin of the History of Medicine*, 5 (1937), pp. 105–14.

57 In his diary Hooke recorded seeing *The Virtuoso* in June 1676, shortly after it opened: "with Godfrey and Tompion at Play, met Oliver there. Damnd Doggs. Vindica me Deus, people almost pointed": cited by Margaret 'Espinasse, *Robert Hooke* (London, 1956), p. 150.

58 Shadwell, *The Virtuoso*, ed. M. H. Nicolson and D. Rhodes (London, 1966), pp. xv ff., 46, 47–8, 55, 102. See also C. Lloyd, "Shadwell and the Virtuosi", *PMLA*, 44 (1929); J. M. Gilde, "Shadwell and the Royal Society: Satire in *The Virtuoso*", *Studies in English Literature, 1500–1900*, 10 (1970), pp. 469–90.

59 See R. S. Westfall, *Never at Rest. A Biography of Isaac Newton* (Cambridge, 1980), pp. 157–8 and notes.

FRANCIS BACON

1 See the "General Preface to the Philosophical Works" by R. L. Ellis, in Bacon's *Works*, ed. Spedding, Ellis and Heath (I, 21–67) for a useful account of Bacon's overall system, and Spedding's preface to the *Parasceve* (I, 367–90) for more detailed comment. Bacon realized that he must issue his *Instauration* in instalments, life not being long enough for him to complete all that he had planned.

2 For Bacon the myth of Proteus was an image of matter, which transformed itself when put under pressure: see *De Sapientia Veterum, The Wisdom of the Ancients*, in *Works*, IV, 725–6.

3 Of the *Novum Organum*: ed. Spedding, *Works*, I, 203, 213–15 (Latin); IV, 95, 106–7 (English).

4 In his collection of *Apophthegmes* (1625) Bacon records how "Vespasian set a tribute [tax] on urine. Titus his son emboldened himself to speak to his father of it: and represented it as a thing indign and sordid. Vespasian said nothing for the time; but a while after, when it was forgotten, sent for a piece of silver out of the tribute money, and called to his son, bidding him smell to it; and asked him, *Whether he found any offence?* Who said, *No. Why lo,* (saith Vespasian again,) *and yet this comes out of urine*" (VII,147).

5 That is, scientific laws can be used to correct errors or discrepancies in experiments: *Works*, I, 213; IV, 105–6.

6 *Works*, I, 142; IV, 29.

7 Bacon's ideas for a scientific research institute first appeared in the early 'devices' of the 1590s: see Spedding's edition, *Works*, VIII, 334–5; XI, 25–6, 65–7 (the *Commentarius Solutus* or notebook of July, 1608), also *The New Atlantis*, ed. G. C. Moore Smith (Cambridge, 1900), pp. x–xvii.

8 This is one of Bacon's favourite Biblical allusions to support his idea for a natural history, from 1 Kings 4:33, "And [Solomon] spake of trees, from the cedar tree that is in Lebanon unto the hyssop that springeth out of the wall", with the idiosyncratic variant of "Mosse" for "hyssop".

9 These form "Tables of Essence and Presence; of Deviation, or of Absence in Proximity; of Degrees, or Comparison": *Novum Organum*, ii, 11–13 (*Works*, IV, 127–37), in the specimen analysis of heat.

10 These represent the "Commencement of Interpretation, or the First Vintage": *Novum Organum*, ii, 20 (IV, 149).

11 Roger Bacon (*c.*1210–1292), scholastic philosopher and Franciscan friar, was traditionally thought to have invented gunpowder.

ROBERT BOYLE

1 The other treatise where Boyle expressed himself "Doubtfully or hesitantly" is presumably *The Sceptical Chymist*.

2 ἐλατήρ: elasticity, expansive power.

3 This analogy is taken from Jean Pecquet, *Experimenta nova anatomica* (Paris, 1651); English translation *New Anatomical Experiments* (London, 1653), chap. 8: excerpts in Marie Boas Hall (ed.), *Nature and Nature's Laws: Documents of the Scientific Revolution* (New York, 1970), pp. 187–202.

4 René Descartes (1596–1650), one of the leading philosophers of the scientific revolution, discussed the nature of air in *Principia Philosophica* (Amsterdam, 1644), Part IV, Principles xlv–xlvii.

5 The "German experiment" was that of Otto Guericke into the vacuum.

6 Boyle reverts to the question of the air's weight in Experiment XXXVI, Works, I, 81–9.

7 Johannes Kepler (1571–1630), a mathematician who made fundamental contributions to astronomy and optics.

8 Giambattista Riccioli (1598–1671), a Jesuit astronomer who made many experiments in the tradition of Galileo.

9 Blaise Pascal (1623–1662), already an outstanding mathematician in his teens, later made experiments to confirm the finding of Evangelista Torricelli that the height of a column of mercury was determined by atmospheric pressure. On 19 September 1648 Pascal caused his brother-in-law François Perier to carry out an experiment which established that the height of the mercury column in a barometer was 67cm in the town of Clermont-Ferrand but 59cm at the top of the Puy-de-Dôme. This showed that atmospheric pressure varied with altitude. Pascal was building on the work of Torricelli (1608–1647), in an experiment performed at Florence in 1643 and recorded in his *Esperienza del argento vivo*. See E. J. Dijksterhuis, *The Mechanization of the World Picture* (Oxford, 1961, 1969), pp. 444–57 for a good treatment of pneumatics; also A. R. Hall, *The Revolution in Science 1500–1700* (London, 1983), pp. 260ff.: "The earliest and perhaps finest example of organized experimental science in the seventeenth century is offered by pneumatics" (p. 200).

10 The "other discourse" Boyle refers to is "Two Essays concerning the Unsuccessfulness of Experiments", included in *Certain Physiological Essays* (London, 1661; *Works*, ed. Birch, I, 298–457).

11 John Wallis (1616–1703), Seth Ward (1617–1689), and Christopher Wren (1632–1723), were all members of the scientific group around John Wilkins at Oxford in the 1650s.

12 Athanasius Kircher (1602–1680), a voluminous Jesuit writer on natural philosophy and magic.

13 Christiaan Huygens (1629–1695), outstanding Dutch mathematician, physicist and astronomer, for twenty years the leading figure in the French Académie royale des sciences. See Dijksterhuis, *The Mechanization of the World Picture*, pp. 368–80, 414, 457–63, 479–80.

14 The use of animals for such experiments was common in this period, as in the famous experiments to determine the role of the lungs in respiration that Hooke performed before the Royal Society in 1663, in which "A Dog was dissected, and by

means of a pair of bellows and a certain Pipe thrust into the Wind-pipe of the Creature, the heart continued beating for a very long while, after all the Thorax and Belly had been open'd...": Sprat, *History of the Royal Society* (1667, p. 232). The experiments proved conclusively that the function of respiration is neither to cool nor to pump (as current theories held) but to supply the lungs with constant fresh air. But although successful, Hooke became disgusted with vivisection, writing to Boyle on 10 November 1664 of this experiment that "I shall hardly, I confess, make [it] again, because it was cruel.... My design was to make some enquiries into the nature of respiration. But... I shall hardly be induced to make any further trials of this kind, because of the torture of the creature..."; Boyle, *Works*, VI, 498. When Hooke was "asked to repeat this operation before the Society, he twice got it postponed and then desired to be excused. But the two doctors who then undertook it were not skilful enough, and the operation 'did not succeed'. So Hooke was finally ordered to do it, and it 'succeeded well'": Margaret 'Espinasse, *Robert Hooke* (London, 1956), p. 52.

15 Galen (129–199 AD), the most considerable doctor and medical writer of antiquity.

16 In this period the word 'chymistry' included alchemy, that is, the manipulation of matter to achieve magical and mystical goals, in accord with an occult or esoteric philosophy. See John Read, *Prelude to Chemistry. An Outline of Alchemy, Its Literature and Relationships* (London, 1936, 1961), still one of the best introductions.

17 Boyle conducts his dialogue through various *personae*, or imaginary characters. *Carneades* was the famous Roman sceptic, one of the leaders of the New Academy, founded by Arcesilas and described in Cicero's *De re publica* (III, vi. 9) and *De finibus* (II, i. 2–3; V, iv. 10). Cicero criticized the negativism of extreme scepticism in morality, but accepted it as a theory of knowledge, in which absolute certainty is impossible, probability the most we can acquire: I would describe Boyle as only provisionally a sceptic. *Eleutherius* means "free-spirited" or generous: he is another of Boyle's mouth-pieces, or at least a sympathetic partner in the dialogue. The two opponents are *Philoponus*, the representative of the alchemists, whose name means "lover of drudgery", and *Themistius*, the Aristotelian: his name means "law-giver".

18 *ubi palam locutus sumus, ibi nihil diximus*: where we speak publicly, we have said nothing.

19 Boyle's work *Some Considerations touching the Usefulnesse of Experimental Philosophy* (Oxford, 1663): vol. 2 (Oxford, 1671); in *Works*, ed. Birch, II, 1–246; III, 392–494.

20 The idea that metals could grow or generate, like organic bodies, was still common at this time: See Bacon, *New Atlantis*, above p. 36.

21 "Peripatetic" (walking about) was a common name for pupils of Aristotle, who is said to have taught in the covered walk (or *peripatos*) of the public gymnasium at Athens known as the Lyceum.

22 Aristotle's logical works were known as the *Organum* or "instrument".

23 Boyle is here satirizing the Aristotelian out of his own mouth, for one of the main charges against that school was that their philosophy was not derived from observation of the real world but imposed on it from some prior theory. Boyle is echoing the technique used by Galileo in his *Dialogue*, of representing Aristotelianism by a comic butt: see Brian Vickers, "Epideictic Rhetoric in Galileo's *Dialogo*", *Annali dell' Istituto e Museo di Storia della Scienze di Firenze*, 8 (1973), pp. 69–102.

24 Theophrastus Philippus Bombastus von Hohenheim (*c*.1493–1541), who called himself Paracelsus, a prolific and influential alchemist.

25 Several works of alchemy were subsequently ascribed to Raymond Lull (*c*.1235–1315), but Lull himself described alchemists as "victims of their own illusions". (I owe this reference to John Melaugh.) On the confusion of alchemical terminology

see Maurice Crosland, *Historical Studies in the Language of Chemistry* (London, 1962).

26 Sulphur being, according to the Paracelsians, the principle and effect of combustion, an "uninflammable sulphur" would be a contradiction in terms.

27 J. B. Van Helmont (1579–1644), alchemist and physician, influenced by Paracelsus but rejecting much of his magical world-view.

28 In its original sense, "contrary to the opinions of most men": see Cicero, *Paradoxa Stoicorum*, Preface.

29 Epicurus (340–270 BC), philosopher of hedonism and atomism. In his system, becoming known through Lucretius' poem *De Rerum Natura*, the gods were conceived as indifferent to the human race, not responsible for their creation, and physical change was also due to purely material causes. Although widely discussed for his atomism, Epicurus remained a dubious figure due to his atheism.

30 "Form" was one of the concepts of Aristotelian physics most vehemently attacked in the scientific Renaissance, for its vagueness and circularity. Boyle delivered a comprehensive refutation in his book *The Origine of Formes and Qualities* (Oxford, 1666): *Works*, ed. Birch, III, 1–137, and in the first of the five *Tracts* (Oxford, 1671); *ibid.*, III, 290–354.

31 1 Kings, 10:22.

HENRY POWER

1 Bacon, *Novum Organum*, ii. 38–9 (*Works*, IV, 192–4).

2 For Descartes' theory that the air is compounded of small "flexible particles" see Boyle above, p. 53.

3 "No beyond", the slogan erected on the pillars of Hercules set in the Straits of Gibraltar, marking the limits of the known world.

4 Boyle, *Certain Physiological Essays* (London, 1661), p. 10.

5 John Tradescant, father and son of that name, collected a famous "cabinet" of rarities which passed on to Elias Ashmole and inspired the building of the Ashmolean Museum, Oxford, 1679–1683.

6 Thomas Moffet or Muffet (1553–1604), author of *The silkeworms and their flies described in verse* (London, 1599), whose work appeared in the collection *Theatrum Insectorum* (London, 1634), English translation *Theatre of Insects* (London, 1658), "the first specialized book on insects" (A. R. Hall, *The Revolution in Science 1500–1700*, p. 149).

7 Proverbs, 6:6.

8 William Harvey (1578–1657) announced his discovery of the circulation of the blood in *De Motu Cordis* (Frankfurt, 1628).

9 The "grazing-Monarch" is presumably King Nebuchadnezzar, who was punished by God, and made to "eat grass as oxen" (Daniel 4:33). *Humanum paucis vivit genus*: humanity is alive in very few people.

10 Another attack on scholastic or Aristotelian philosophy.

11 As Marie Boas Hall glosses this passage: "These are all examples of scientific follies or failures. The alkahest is the universal solvent and the philosophers' stone the universal catalyst; there was then no means of determining longitude accurately, though the correct principles were known; squaring the circle is finding an area bounded by straight lines equal to the area of the circle – an impossibility; the problem of the trisection of an angle was to trisect it by using ruler and compass only. Similarly, the problems mentioned in the next sentence were then incompletely resolved, though great discoveries were to be made in the next twenty-five years." *Nature and Nature's Laws. Documents of the Scientific Revolution*, ed. M. B. Hall (New York, 1970), p. 128 note.

12 George Hakewill (1578–1649), author of *An apologie or declaration of the power and providence of God in the government of the world* (Oxford, 1627), which attacked the notion that the universe is decaying. See Victor Harris, *All Coherence Gone* (Chicago, 1949).

13 *nil dictum, quod non prius dictum*: nothing can be said that has not been said before in earlier times.

14 Hippocrates (*c.*460–380 BC), Greek doctor and medical writer, whose *Aphorisms* begin with the famous phrase *ars longa, vita brevis est* ("the life so short, the craft so long to learn", as Chaucer's Wife of Bath has it).

15 For Boyle see above, section 3.

16 An allusion to Sir Thomas Browne, *Religio Medici* (London, 1643), I. 9: "Methinks there be not impossibilities enough in Religion for an active faith."

17 Democritus (*c.*460–370 BC) adopted the atomic theory of his master Leucippus. According to an ancient tradition he held that truth lives at the bottom of a deep well.

18 Another name for Aristotle, who was born at Stagirus (later Stagira) in Chalcidice.

19 Xerxes, King of Persia 485–465 BC. In his attack on Greece he bridged the Hellespont and dug a canal through the Athos peninsula.

ROBERT HOOKE

1 Many of these ideas come from Bacon: see, e.g., *Works*, IV, 26–7, 58, 164–5, etc.

2 Hooke follows Bacon and Boyle in attacking the abstractions of Aristotelian philosophy.

3 Such celebrations of modern scientific discoveries were conventional in the literature defending the new sciences, such as Sprat's *History of the Royal Society* and Glanvill's *Plus Ultra*.

4 Similar tributes were made to Boyle by Glanvill, *Plus Ultra* (London, 1668) in the section "Modern Improvements of Useful Knowledge", pp. 92–107; and to Christopher Wren by Sprat, *History of the Royal Society* (London, 1667), pp. 311–18. The new sciences were already mythologizing their heroes.

5 See Power, above p. 91.

6 Hooke had described his method of focusing additional light on the object to be examined.

7 Occult qualities, according to Aristotelian science, were the hidden or invisible properties of things which affected their existence or behaviour.

8 A reference to Bacon's distinction between "experiments of light", advancing theoretical knowledge, and "experiments of fruit", yielding practical results: *Works*, IV, 70–1, 95.

9 Presumably Wren's drawings from microscopical enlargements had passed into the collection of Charles II.

10 Bacon, *Novum Organum*, i. 64; *Works*, IV, 65.

11 *forma informans*: the technical term in Aristotelian philosophy for "that which makes anything (matter) a determinate species or kind of being" (*OED*).

12 This is the first use of the term "cell" in its biological sense.

13 *intus existens*: something existing inside.

14 Hooke corrects Power's supposition, p. 91 above.

15 It was the tenet of Epicureanism that the created universe was produced not by the deity but by a fortuitous concourse of atoms.

16 Hooke's Observations XLVII "*Of the* Shepherd Spider" and XLVIII "*Of the* hunting Spider *and several other sorts of* Spiders" gave rise to Shadwell's mockery in *The*

Virtuoso, edition cited, I, ii, 7 ff. and III, iii, 40 ff., where the butt Sir Nicholas Gimcrack is made to expatiate on the "six and thirty several sorts of spiders" in England, "your hound, greyhound, lurcher, spaniel, spider", and even "a tame spider call'd Nick ... the best natur'd, best condition'd spider that ever I met with ... of the spaniel breed, sir".

17 Richard Ligon, a traveller, published *A true and exact history of the island of Barbados* (London, 1657).

18 *tamquam animam expirasset*: as if he had expired.

19 Bacon, *Novum Organum*, ii. 39; *Works*, IV, 193–4.

20 Compare Bacon in the *Parasceve*, above, p. 22, 29; and the *New Atlantis*, pp. 42–3.

21 Cf. Bacon, *Parasceve*, above, p. 25.

22 Jacques Gaffarel (1601–1681), librarian to Cardinal Richelieu, author of numerous occult works drawing on the Kabbalah.

23 The *Natural History*, in 36 Books, of Pliny the Elder (24–79 AD) remained a respected source of information in the Renaissance, despite containing much "marvellous" matter.

24 Olearius: Adam Olearius (1591?–1671), who travelled extensively in Russia and the Far East. See *The Voyages & travels of the ambassadors sent by Frederick duke of Holstein, to the great Duke of Muscovy, and the King of Persia ...*, *written originally by Adam Olearius*, tr. J. Davies (London, 1662). Boyle quotes frequently from the French translation, *Voyages de Moscovie & de Perse* (Paris, 1656).

25 See Genesis, 3:2–7.

26 Seneca, *Naturales Quaestiones* III, 27. 3: "Any deviaton by nature from the existing state of the universe is enough for the destruction of mankind. So, when that destined time comes the fates put into motion many causes at the same time. For according to some thinkers, among them Fabianus, such a great change does not occur without a shattering of the universe."

27 The *Gigantomachia* or war of the giants against the gods was one of the most popular myths in Greece: see Apollodorus, I, 34 ff.

THOMAS SPRAT

1 Bacon was a lawyer and politician. A Member of Parliament from 1584 to 1621, his public career, blocked by Queen Elizabeth in 1593 on account of his opposition to her tax demands, flourished under James I. He became King's Counsel in 1604, Solicitor-General in 1607, Attorney-General in 1613, a Privy Councillor in 1616, Lord Keeper in 1617, Lord Chancellor in 1618; he was created Baron Verulam in 1618, and Viscount St. Albans in 1621. In that year he was dismissed from all offices for having taken presents from two men whose cases he had tried in court. This was a common practice at the time, but Bacon did not allow it to affect his judgment. The remaining five years of his life were devoted to research and writing.

2 In the Civil War the word "committee" had been associated with the Parliamentary, rebellious party.

3 Compare Bacon: "For the mind of man is far from the nature of a clear and equal glass, wherein the beams of things should reflect according to their true incidence; nay it is rather like an enchanted glass, full of superstition and imposture, if it be not delivered and reduced" (III, 394–5).

4 Compare Bacon's *Parasceve*, above p. 25.

5 Having threatened to emulate Plato, who wanted the poets to be expelled from his *Republic*, Sprat now echoes Aristotle, who pronounced it lawful for man to cultivate "speech and reason" in order to defend himself: *Rhetorica*, 1355b 1–7.

6 Compare Bacon: above, p. 24.

7 Bacon, *Works*, IV, 17, 71, 95–6.
8 Bacon, *Works*, IV, 62–3.
9 Alluding to the proverb *homo homini lupus*, "a man is a wolf to another man" (from Plautus, *Asinaria*, 495).
10 Genesis 49:21.
11 Genesis 49:22.

JOHN WILKINS

1 Perhaps Piso the friend of Galen: see Philipp Labbe, *Claudii Galeni chronologicum elogium* (1660), and Pauly-Wissowa, *Real-Encyclopädie der Classischen Altertumswissenschaft*, Band XX. 2 (Stuttgart, 1950), cols. 1,802–3.
2 Bacon's views on the "real character" can be found in *Works*, III, 121; IV, 39–40. Gerard Johannes Vossius (1577–1649), author of *De arte grammatica* (Amsterdam, 1635) – a work of over 1,200 pages (also known as the *Aristarchus*) – and *Latina syntaxis* (Leiden, 1631).
3 *scriptio est vocum pictura*: writing is the image of sounds.
4 Julius Caesar Scaliger (1484–1588), author of *De causis linguae latinae* (Lyons, 1540). On the Renaissance grammarians cited by Wilkins see G. A. Padley, *Grammatical Theory in Western Europe 1500–1700. The Latin Tradition* (Cambridge, 1976).
5 *scientia loquendi ex usu*: the knowledge of speaking from experience.
6 Bacon, *De Augmentis Scientiarum*, Book VI, chap. 1: *Works*, IV, 439–40. In his English version of *De Augmentis*, *Of the Advancement and Proficience of Learning* (Oxford, 1640), Gilbert Watts translates this section of the *Desiderata* or "Catalogue of *Deficients*" as "A Philosophicall Grammar, Or the Analogy between Words & Things" (sig. Pppp₂ʳ).
7 Marcus Terentius Varro, the greatest Roman scholar (116–27 BC), wrote *De lingua Latina libri XXV*, of which two complete and four fragmentary books remain.

ISAAC NEWTON

1 In fact, Newton's notebooks reveal that he started working with prisms in 1664–5. See R. S. Westfall, *Never at Rest. A Biography of Isaac Newton* (Cambridge, 1980), pp. 156 ff., with the literature cited there.
2 In his edition of Newton's *Correspondence*, vol. 1 (Cambridge, 1959), p. 103, H. W. Turnbull notes that Newton may be referring to an experiment with a prism described by Descartes in *Les Météores*, designed to account for "the formation and the colours of the rainbow. Sunlight, falling almost perpendicularly upon the slanting face of a prism and passing through it, is refracted out of the furthest lowest and horizontal face at a slit made in an opaque support, and causes all the colours of the rainbow to appear below on a vertical screen, from red to blue".
3 The sine-law ($\sin i/\sin r = k$) was discovered by Willebrord Snel (1591–1626), but first published as an independent discovery by Descartes in *La Dioptrique* (1637).
4 That is, the sum of 63°12′ and 44°56′ is twice 54°4′, the angle of incidence on the prism.
5 In transcribing Newton's letter Oldenburg changed this passage, which originally read "my own & others Experience".
6 In those days tennis ("real" or royal tennis, as it survives), was played in an indoor court, rather like squash rackets today. Newton's analysis of the causes of a ball swerving is remarkably accurate.
7 This is Descartes' theory of light, also discussed by Hooke in *Micrographia* (1665), p. 54.

8 Bacon conceived the *instantia crucis* in *Novum Organum*, II. xxxvi (not xxxv, as in Turnbull) as one of his "Prerogative instances" for evaluating the findings of an experiment. This "crucial instance", or "instance of the fingerpost", as Bacon explains, "borrows the term from the fingerposts [or 'crosses'] which are set up where roads part, to indicate the several directions" (*Works*, IV, 180), and refers to the decisive experiment devised to distinguish the true explanation from the false. The post-Baconian form, "*experimentum crucis*" was coined by Boyle (not Hooke, as usually stated), in his *Defence of the Doctrine touching the Spring and Weight of the Air* (1662), referring to Pascal's experiment on the Puy-de-Dôme as "an *experimentum crucis* (to speak with our illustrious *Verulam*)": *Works*, I, 151. Hooke adopted the phrase in *Micrographia* (1665), p. 54. See also J. A. Lohne, "Experimentum Crucis", *Notes and Records of the Royal Society*, 23 (1968), pp. 169–99.

9 Newton left Cambridge in June 1666, returning in 1668.

10 There are references to a 4 foot tube in Newton's earlier letters, and one of 6 feet was described at the Royal Society in February 1672: Turnbull, p. 104.

11 Oldenburg omitted from the *Transactions* the following passage: "A naturalist would scearce expect to see ye science of those [colours] become mathematicall, & yet I dare affirm that there is as much certainty in it as in any other part of Opticks. For what I shall tell concerning them is not an Hypothesis but most rigid consequence, not conjectured by barely inferring 'tis thus because not otherwise or because it satisfies all phaenomena (the Philosophers universall Topick), but evinced by ye mediation of experiments concluding directly & without any suspicion of doubt. To continue the historicall [factual] narration of these experiments would make a discourse too tedious & confused, & therefore I shall rather lay down the *Doctrine* first, and then, for its examination, give you an instance or two of the *Experiments* as a specimen of the rest": *Correspondence*, I, 96–7.

12 "*Lignum Nephriticum* is the wood of a Central American shrub or small tree ... which, when infused in slightly acidulated water, yields a yellow solution which also gives a blue light by luminescence": A. R. and M. B. Hall (eds.), *The Correspondence of Henry Oldenburg*, IX (Madison, Wis., 1973), p. 28.

13 Hooke, *Micrographia*, p. 73. Newton made extensive notes on this book (printed in Keynes's *Bibliography* of Hooke, pp. 97 ff.), his disagreement with Hooke's explanation of light provoking his own theoretical work.

14 Turnbull cites an early printed form of this letter in which Newton had added various footnotes glossing his meaning. Here he notes: "Through an improper distinction which some make of mechanical Hypotheses into those where light is put [said to be] a body and those where it is put the action of a body, understanding the first of bodies trajected through a medium, the last of motion or pression propagated through it, this place [passage] may be by some unwarily understood of the former. Whereas light is equally a body or the action of a body in both cases. If you call its rays the bodies trajected in the former case, then in the latter case they are the bodies which propagate motion from one to another in right [straight] lines till the last strike the sense. The only difference is that in one case a ray is but one body, in the other many. So in the latter case, if you call the rays motions propagated through bodies, in the former it will be motions continued in the same bodies. The bodies in both cases must cause vision by their motion.

Now in this place my design being to touch upon [allude to] the notion of the Peripateticks, I took not body in opposition to motion, as in the said distinction, but in opposition to a Peripatetick quality, stating the question between the Peripateticks and Mechanick Philosophy by inquiring whether light be a quality or body. Which that it was my meaning may appear by my joyning this question with others hitherto disputed between the two Philosophies; and using in respect of one side the Peripatetick terms *Quality, Subject, Substance, Sensible qualities*; in respect of the other the Mechanick ones *Body, Modes, Actions*; and leaving undetermined the Kinds of those actions (suppose whether they be pressions, strokes, or other dashings), by which light may produce in our minds the phantasms of colours." *Correspondence* I, 105–6.

15 Newton added in a note: "understand therefore these expressions to be used here in respect of the Peripatetick Philosophy. For I do not my self esteem colours the qualities of light, or of anything else without us, but modes of sensation excited in our minds by light. Yet because they are generally attributed to things without us, to comply in some measure with this notion I have in other places of these letters attributed them to the rays rather than to bodies, calling the rays for their effect on the sense, red, yellow, &c. whereas they might be more properly called rubriform, flaviform, &c.": *Correspondence* I, 106.

16 Although Oldenburg informed Newton that his letter had been read to the Royal Society and received "both with a singular attention and an uncommon applause" (*Correspondence* I, 107), several contemporaries, ranging from lesser figures such as Father Ignatius Pardies to outstanding scientists such as Hooke and Huygens, could not bring themselves to accept Newton's theory, and he became involved in some lengthy controversies. See his *Correspondence*, vol. I, *passim*; I. B. Cohen (ed.), *Isaac Newton's Papers & Letters on Natural Philosophy*, second ed. (Cambridge, Mass., 1978), pp. 27–238, 505–22; and *The Correspondence of Henry Oldenburg*, ed. A. R. and M. B. Hall (Madison, Wis.), vols. VIII (1971) and IX (1973), covering the period 1671–1673. Newton was surprised at the reluctance to accept his theory that white light is merely a combination of all the colours, writing in 1675, "This I believe hath seemed the most Paradoxicall of all my assertions, & met with the most universall & obstinate Prejudice. But to me it appears as infallibly true & certaine as it can seem extravagant to others": *Correspondence* I, 385. We should remember the original meaning of "paradox": "contrary to the opinion of most men". Newton restated his theory in answering Pardies, Hooke, Huygens, and Oldenburg, with some added refinements. But the main effect the controversy had on him was to make him suspicious of further publication, since it might lead to more protracted disputes.

APPENDIX

1 The anecdote of Apelles (4th c. BC), the painter who one day, dissatisfied with his progress, threw his sponge at the painting and achieved by chance what he could not manage by design, was widely disseminated.

2 In Boyle, *Works*, ed. T. Birch, I, 298–457.

3 See, e.g., *Works*, I, 491.

4 In *Works*, II, 462–734.

5 Stephen Medcalf (ed.), *The Vanity of Dogmatizing* (Hove, Sussex, 1970), p. xxxiv. This useful edition reprints the three versions in facsimile; further references in the text will be to the pagination of the originals.

Select Bibliography

1. Reference

For biographical and critical accounts, together with bibliographies, see G. C. Gillespie (ed.), *Dictionary of Scientific Biography*, 16 vols. (New York, 1970–1980). A useful bibliography of 17th-century science is provided by Charles Webster in *The New Cambridge Bibliography of English Literature, vol. I, 600–1660*, ed. G. Watson (Cambridge, 1974), cols. 2,343–80. A rather specialized bibliography, which does include discussion of Bacon, Wilkins, and universal language schemes, is Herbert E. Brekle, "The Seventeenth Century", in *Current Trends in Linguistics*, 13 (1975), pp. 277–382.

2. Science: General

Three classic introductory studies of the scientific revolution are E. A. Burtt, *The metaphysical foundations of modern physical science* (London, 1924; paperback); Alexandre Koyré, *From the Closed World to the Infinite Universe* (Baltimore, Md., 1957; paperback), and E. J. Dijksterhuis, *The Mechanization of the World Picture*, tr. C. Dikshoorn (Oxford, 1961; paperback). Also useful are R. S. Westfall, *The Construction of Modern Science. Mechanisms and Mechanics* (Cambridge, 1977); A. R. Hall, *The Revolution in Science 1500–1750* (London, 1983), and Michael Hunter, *Science and Society in Restoration England* (Cambridge, 1981) with a valuable "Bibliographical essay", pp. 198–219. Charles Webster's study, *The Great Instauration: Science, Medicine and Reform 1626–1660* (London, 1975) is a massive and superbly documented study of the influence of Baconian science on the Hartlib circle and associated movements during the Civil War and Interregnum periods, but too heavily biassed towards "Puritan" science as a major enterprise (most of their projects for social and educational reform were well meaning but unrealized). Valuable studies of medical research are R. G. Frank, jr., *Harvey and the Oxford physiologists: a study of scientific ideas and social interaction* (Berkeley and Los Angeles, Ca., 1980), and of the treatment of mental illness, Michael MacDonald, *Mystical Bedlam. Madness, anxiety, and healing in seventeenth-century England* (Cambridge, 1981).

Two useful anthologies are Norman Davy (ed.), *British Scientific Literature in the Seventeenth Century* (London, 1953), and Marie Boas Hall (ed.), *Nature and Nature's Laws. Documents of the Scientific Revolution* (New York, 1970) which covers the whole of Europe from Copernicus to Newton, with valuable introductions.

Select bibliography

3. Science: Individual authors

Bacon: The starting point for all Bacon studies is still *The Works of Francis Bacon*, ed. James Spedding, R. L. Ellis, and D. D. Heath, 14 vols. (London, 1857–1874; facsimile reprints Stuttgart, 1962 and New York, 1968). This includes both English works and Latin texts with translations. Three minor Latin texts not translated by Spedding were rendered into English by Benjamin Farrington in *The Philosophy of Francis Bacon* (Liverpool, 1964). Some newly discovered manuscript material of early and unpublished drafts is presented by Graham Rees, assisted by Christopher Upton, *Francis Bacon's Natural Philosophy: A New Source* (Chalfont St. Giles, 1984). On Bacon as a scientist see Paolo Rossi, *Francis Bacon. From Magic to Science*, tr. S. Rabinovitch (London, 1968); and essays by Mary Hesse and M. E. Prior in Brian Vickers (ed.), *Essential Articles for the Study of Francis Bacon* (Hamden, Conn., 1968; London, 1972); Mary Horton, "In Defence of Francis Bacon. A Criticism of the Critics of the Inductive Method", *Studies in the History and Philosophy of Science*, 4 (1973), pp. 241–78. The best recent study of *New Atlantis* is by J. C. Davis, in *Utopia & the ideal society. A study of English utopian writing 1516–1700* (Cambridge, 1981), pp. 105–37.

Boyle: The standard biography is R. E. W. Maddison, *The Life of the Hon. Robert Boyle, F.R.S.* (London, 1961). The major edition is *The Works of the Honourable Robert Boyle*, ed. Thomas Birch, 5 vols. (London, 1744); rev. ed. 6 vols. (London, 1772; facsimile reprint, Hildesheim, 1969). J. Fulton, *A Bibliography of the Honourable Robert Boyle*, 2nd ed. (Oxford, 1961) includes both primary literature and secondary literature up to 1960. Marie Boas Hall has edited selections in *Robert Boyle on Natural Philosophy* (Bloomington, Ind., 1965). Another valuable anthology is M. A. Stewart (ed.), *Selected Philosophical Papers of Robert Boyle* (Manchester, 1979), which includes material published for the first time, and lists textual variants (the need for a new edition of Boyle becomes greater every year). *The Sceptical Chymist* is available in several modern editions, most conveniently in the Everyman Library, ed. E. A. Moelwyn-Hughes (London, 1964). On Boyle as a scientist see Marie Boas Hall, *Robert Boyle and Seventeenth-Century Chemistry* (Cambridge, 1958; New York, 1968); *idem*, "The Establishment of the Mechanical Philosophy", *Osiris*, 10 (1952), pp. 412–541; Thomas S. Kuhn, "Robert Boyle and Structural Chemistry in the Seventeenth Century", *Isis*, 43 (1952), pp. 12–36; Maurice Mandelbaum, *Philosophy, Science and Sense Perception* (Baltimore, Md., 1964), on his links with Locke; E. M. Klaaren, *Religious Origins of Modern Science* (Grand Rapids, Mi., 1977); R. S. Westfall, "Unpublished Boyle papers relating to scientific method", *Annals of Science*, 12 (1956), pp. 63–73, 103–17; J. E. McGuire, "Boyle's conception of nature", *Journal of the History of Ideas*, 33 (1972), pp. 523–42; Simon Schaffer and Steven Shapin, *Leviathan and the air-pump: Hobbes, Boyle, and the experimental life* (Princeton, NJ, 1985). J. R. Jacob, *Robert Boyle and the English Revolution* (New York, 1977), puts Boyle in the immediate context of the Civil War but cuts him off from other ethical and philosophical traditions.

Select bibliography

Power: *Experimental Philosophy* has been reprinted with an important historical and biographical introduction by Marie Boas Hall (New York, 1966). Power's correspondence with Browne can be found in Geoffrey Keynes (ed.), *The Works of Sir Thomas Browne*, 4 vols., 2nd ed. (London, 1964), IV, 254–70. See also T. Cowles, "Dr. Henry Power, Disciple of Sir Thomas Browne", *Isis*, 20 (1934), pp. 344–66; Charles Webster, *The Great Instauration*, and *idem*, "Henry Power's experimental philosophy", *Ambix*, 14 (1967), pp. 150–78, which seems to me to overstate Power's links with occult science.

Hooke: The only biography is Margaret 'Espinasse, *Robert Hooke* (London, 1956) but it overvalues Hooke and downgrades Newton. R. T. Gunther's collection, *Early Science in Oxford*, 14 vols. (Oxford, 1923–1945), devotes five volumes to Hooke's life and works: vols. VI, VII (extracts from contemporary writers), VIII (the *Lectiones Cutlerianae*), X (Hooke's first work, on capillary phenomena, and his diary for 1688–1693), and XIII (*Micrographia*; repr., New York, 1961). H. W. Robinson and W. Adams edited *The Diary of Robert Hooke, M.A., M.D., F.R.S., 1672–1680* (London, 1935). *The Posthumous Works of Robert Hooke*, ed. Richard Waller (London, 1705), is available in two modern facsimile reprints with helpful introductions, ed. R. S. Westfall (New York, 1969), and Theodore Brown (London, 1971). William Derham included further unpublished work in *Philosophical Experiments and Observations of the Late Eminent Dr. Robert Hooke ... and Other Eminent Virtuoso's in His Time* (London, 1726). Geoffrey Keynes compiled *A Bibliography of Dr. Robert Hooke* (Oxford, 1960). On Hooke as a scientist see E. N. da C. Andrade, "Robert Hooke", *Proceedings of the Royal Society*, 201A (1950), pp. 439–73; Mary Hesse, "Hooke's Philosophical Algebra", *Isis*, 57 (1966), pp. 67–83; *idem*, "Hooke's Vibration Theory and the Isochrony of Springs", *ibid.*, pp. 433–41; Alexandre Koyré, "An Unpublished Letter of Robert Hooke to Isaac Newton", *Isis*, 43 (1952), pp. 312–37, republished in Koyré, *Newton Studies* (Cambridge, Mass., 1965); R. S. Westfall, "Hooke and the Law of Universal Gravitation", *British Journal of the History of Science*, 3 (1967), pp. 245–61; A. I. Sabra, *Theories of Light from Descartes to Newton* (London, 1967); A. P. Rossiter, "The First English Geologist", *Durham University Journal*, 27 (1935), pp. 172–81; D. J. Holroyd, "Robert Hooke's methodology of science as exemplified by his 'Discourse of earthquakes'", *British Journal for the History of Science*, 6 (1972), pp. 109–30; F. F. Centore, *Robert Hooke's contributions to mechanics* (The Hague, 1970).

Sprat and the Royal Society: A facsimile edition of *The History of the Royal Society* was edited by J. I. Cope and H. W. Jones (St. Louis, Mo., and London, 1958, 1966), with a somewhat unhelpful introduction. Much information on Sprat is contained in Thomas Birch, *The History of the Royal Society*, 4 vols. (London, 1756–57), also available in a modern facsimile, ed. A. R. and M. B. Hall (New York, 1968). On Sprat's Baconianism see H. Fisch and H. W. Jones, "Bacon's Influence on Sprat's *History of the Royal Society*", *Modern Language Quarterly*, 12 (1951), pp. 399–406; on his prose style see Robert Cluett, "Style, Precept, Personality: A Test Case (Thomas Sprat,

1635–1713)", *Computers and the Humanities*, 5 (1970–71), pp. 257–77, drawing on his 1969 Columbia Ph.D. thesis "These Seeming Mysteries: The Mind and Style of Thomas Sprat", *Dissertation Abstracts*, 70, p. 6,950. For Sprat's role as a propagandist and defender of the Royal Society against its enemies see Paul B. Wood, "Methodology and Apologetics: Thomas Sprat's *History of the Royal Society*", *British Journal for the History of Science*, 13 (1980), pp. 1–26, and the essay by Brian Vickers listed under "Prose Style", below.

On the Royal Society the older studies by C. R. Weld, *A history of the Royal Society* (London, 1848); Henry Lyons, *The Royal Society 1660–1940* (Cambridge, 1940); and Dorothy Stimson, *Scientists and Amateurs. A History of the Royal Society* (New York, 1948) are still useful, while a valuable series of essays on its founding members was edited by Harold Hartley, *The Royal Society: its origins and founders* (London, 1960). By contrast, one modern study, Margery Purver, *The Royal Society: Concept and Creation* (London, 1967), makes the mistake of taking Sprat as gospel truth: see Charles Webster's critical review, "The Origins of the Royal Society", *History of Science*, 6 (1967), pp. 106–28. A fresh start is provided by Michael Hunter, *The Royal Society and its Fellows 1660–1700: The Morphology of an Early Scientific Institution* (Chalfont St. Giles, 1982), expanded from an earlier study in *Notes and Records of the Royal Society of London*, 31 (1976), pp. 9–114. See also K. T. Hoppen, *The common scientist in the seventeenth century* (London, 1970), and *idem*, "The nature of the early Royal Society", *British Journal for the History of Science*, 9 (1976), pp. 1–24, 243–73; R. G. Frank, jr., "Institutional structure and scientific activity in the early Royal Society", *Proceedings of the fourteenth congress of the history of science* (1974) (Tokyo, 1975), vol. IV, pp. 82–101; Walter E. Houghton, "The English virtuoso in the seventeenth century", *Journal of the History of Ideas*, 3 (1942), pp. 51–73, 190–219; and *idem*, "The History of Trades: its relation to seventeenth-century thought", *ibid.*, 2 (1941), pp. 33–60; Margaret Denny, "The early program of the Royal Society and John Evelyn", *Modern Language Quarterly*, 1 (1940), pp. 481–97.

Wilkins: The book-length study by Barbara Shapiro, *John Wilkins 1614–72: an intellectual biography* (Berkeley and Los Angeles, Ca., 1969), is somewhat disappointing. A more incisive account is the article on Wilkins by Hans Aarsleff in *Dictionary of Scientific Biography*, vol. XIV, pp. 361–81, reprinted in Aarsleff, *From Locke to Saussure* (London, 1982). The *Essay Towards a Real Character and a Philosophical Language* was reprinted in facsimile (Menston, Yorks., 1968). Most of Wilkins's other writings were reprinted in *The Mathematical and Philosophical Works* (London, 1708, 1802). On the controversy with John Webster, in which Wilkins was co-author with Seth Ward of *Vindiciae Academiarum*, see A. Debus (ed.), *Science and education in the seventeenth century: the Webster–Ward debate* (London, 1970). See Mary Slaughter, *Universal Languages and Scientific Taxonomy in the Seventeenth Century* (Cambridge, 1982); Vivian Salmon, *The Study of Language in Seventeenth-Century England* (Amsterdam, 1979); Clark Emery, "John Wilkins' Universal Language", *Isis*, 38 (1947–8), pp. 174–85. On Wilkins's theology a useful study is H. R. McAdoo, *The Spirit of Anglicanism. A Survey of Anglican Theological Method in the Seventeenth Century* (London, 1965),

Select bibliography

which surprisingly omits consideration of Wilkins's *Ecclesiastes, or a Discourse Concerning the Gift of Preaching* (London, 1646); B. C. Vickery, "The Significance of John Wilkins in the History of Bibliographical Classification", *Libri*, 2 (1953), pp. 326–43.

Newton: P. and R. Wallis have issued a detailed bibliography, *Newton and Newtonia, 1672–1975* (Folkestone, 1977). R. S. Westfall, *Never at Rest: A Biography of Isaac Newton* (Cambridge, 1980), is a remarkable synthesis of a vast amount of material, printed and manuscript, as well as a very good exposition of Newton's science. For a shorter but wide-ranging account, with an excellent bibliography, see the article on Newton by I. B. Cohen in *Dictionary of Scientific Biography*, vol. x, pp. 42–101. The major works of Newton have received editorial treatment of the highest quality. His early essays on optics are reprinted in facsimile, with a valuable introduction by Thomas S. Kuhn, in I. B. Cohen (ed.), *Isaac Newton's Papers & Letters on Natural Philosophy*, 2nd ed. (Cambridge, Mass., 1978). The *Opticks* is available in reprint, e.g. ed. D. H. D. Roller (New York, 1952); an edition was promised by the late Henry Guerlac. The most rigorous edition of the *Principia* is by A. Koyré, I. B. Cohen and Anne Whitman, *Isaac Newton's Philosophiae naturalis principia mathematica, the Third Edition (1726) With Variant Readings*, 2 vols. (Cambridge, 1972). The standard translation is *Mathematical Principles of Natural Philosophy*, by Andrew Motte, rev. Florian Cajori (Berkeley, Ca., 1934), although in need of correction. A. R. and M. B. Hall have edited *Unpublished Scientific Papers of Isaac Newton* (Cambridge, 1964). His *Mathematical Papers* have been edited by D. T. Whiteside, 8 vols. (Cambridge, 1959–1981); his *Optical Papers* by A. E. Shapiro (vol. i: *The optical lectures 1670–1672*, Cambridge, 1984); his *Correspondence* by H. W. Turnbull, J. F. Scott, A. R. Hall, and Laura Tilling, 7 vols. (Cambridge, 1959–1978), and there are plans to publish the manuscripts on alchemy and biblical chronology.

Of the vast literature on Newton's scientific achievements see R. S. Westfall, *Force in Newton's Physics. The Science of Dynamics in the Seventeenth Century* (London, 1971); A. I. Sabra, *Theories of Light from Descartes to Newton* (London, 1967); I. B. Cohen, *The Newtonian Revolution. With Illustrations of the Transformation of Scientific Ideas* (Cambridge, 1981); Alexandre Koyré, *Newtonian Studies* (Cambridge, Mass., 1965), also available in a corrected French translation, *Études newtoniennes* (Paris, 1968); Robert Palter (ed.), *The 'Annus Mirabilis' of Sir Isaac Newton, 1666–1966* (Cambridge, Mass., 1970), a useful collection of essays on scientific, philosophical, and religious topics; Frank E. Manual, *Isaac Newton, Historian* (Cambridge, Mass., 1963); Marjorie Hope Nicolson, *Newton Demands the Muse* (Princeton, NJ, 1946); B. J. Dobbs, *The Foundations of Newton's Alchemy: The Hunting of the Greene Lyon* (Cambridge, 1975).

4. Prose style

There is no adequate study of English prose in this period. Two earlier writers whose work exercised influence, but should now be used with great caution, are Morris W. Croll, in essays dating from 1914 to 1929, collected in

Select bibliography

Style, Rhetoric, and Rhythm, ed. J. M. Patrick *et al.* (Princeton NJ, 1966), and R. F. Jones, in essays dating from 1920 to 1940 collected in *The Seventeenth Century*, ed. Jones *et al.* (Stanford, Ca., 1951). Croll discussed prose in terms of categories loosely derived from Greek literary theory ("Attic" and "Asiatic") or Renaissance philology ("Ciceronian", "Anti-Ciceronian", "Senecan"), and then classified writers or schools as belonging to one or another "movement" on the basis of brief quotations and little actual analysis. His unhistorical and impressionistic methods were taken to a self-defeating extreme by George Williamson, *The Senecan Amble: A Study in Prose Form from Bacon to Collier* (London, 1951). On the defects of this approach see Robert Adolph, *The Rise of Modern Prose Style* (Cambridge, Mass., 1968), and Brian Vickers, *Francis Bacon and Renaissance Prose* (Cambridge, 1968), pp. 96–140.

R. F. Jones worked more closely with primary texts, yet took Sprat's description of the Royal Society's "reform" of prose style as an historical fact, and went on to describe the scientific movement as sharing an antipathy to ornateness, rhetoric, metaphor, and even language itself, their ruthless search for "utility" and "materialism" resulting in a hard, "thing-like" approach to literature and life. An equally simplistic account of this supposed revolution was given by Francis Christensen, "John Wilkins and the Royal Society's Reform of Prose Style", *Modern Language Quarterly*, 7 (1946), pp. 179–87, 279–90. For objections to such reading see William Youngren, "Generality, Science, and Poetic Language in the Restoration", *ELH*, 35 (1968), pp. 158–87, who argues that post-Restoration literature does not reflect a mechanistic, colourless universe; and Brian Vickers, "The Royal Society and English Prose Style: A Reassessment", in *Rhetoric and the Pursuit of Truth: Language Change in the Seventeenth and Eighteenth Centuries*. Seminars held at The Clark Library, Los Angeles, March 1980 (Los Angeles, Ca., 1985), pp. 1–76. The fundamental research on this topic has yet to be done.